Acoustic Therapy Devices

Analyzing physiological bodily signals for medical patient detection

Elmo Contreras

Table of Contents

1 Chapter 1: Introduction to Acoustic Therapy

Acoustic therapy, at its core, represents a non-invasive therapeutic approach leveraging sound waves to stimulate physiological changes within the body. It's not merely about listening to pleasant melodies; it's a sophisticated method that utilizes precisely calibrated sound frequencies and intensities to influence cellular activity, nervous system function, and overall physiological homeostasis. This approach distinguishes it from other sound-based practices like music therapy, which primarily focuses on emotional and psychological well-being. Acoustic therapy seeks direct physical effects. Think of it as a targeted intervention rather than a generalized relaxation technique, although relaxation might be a beneficial side effect.

The historical roots of acoustic therapy are surprisingly deep. Ancient cultures, from Aboriginal Australians using didgeridoos for healing ceremonies to Tibetan monks employing singing bowls for meditation and therapeutic purposes, intuitively understood the power of sound. These traditions, passed down through generations, suggest an inherent human understanding of sound's capacity to affect physical and mental states. However, modern acoustic therapy, as we understand it today, is firmly grounded in scientific principles. The advent of sophisticated recording equipment, signal processing techniques, and advanced understanding of human physiology has allowed us to move beyond anecdotal evidence and explore the precise mechanisms by which sound can induce therapeutic effects. Early research, particularly in the fields of sonochemistry and sonoluminescence, demonstrated the ability of sound waves to interact with matter at a molecular level. This paved the way for exploring its potential in biological systems.

From these foundations, diverse applications of acoustic therapy have emerged, ranging from pain management to tissue regeneration. Focused ultrasound, for example, is used to break up kidney stones and deliver drugs directly to tumors. Low-intensity pulsed ultrasound (LIPUS) is employed to accelerate bone fracture healing. In the realm of mental health, binaural beats and isochronic tones are being investigated for their potential to reduce anxiety and improve sleep quality. What ties these diverse applications together is the underlying principle of using sound as a targeted stimulus to elicit a specific physiological response. My fascination started with this, when my daughter was diagnosed with severe panic attacks, and I couldn't sleep properly.

Physiological signal analysis is the keystone to understanding and optimizing acoustic therapy protocols. Without objective measurements of the body's response to sound, we're left with subjective reports and guesswork. Physiological signals, such as heart rate variability (HRV), electroencephalography (EEG), electromyography (EMG), and respiratory rate, provide a window into the body's internal state. These signals can be acquired non-invasively using a variety of sensors, and their analysis can reveal how the body is responding to acoustic stimulation. Analyzing a patient's heart sounds, for instance, can help you diagnose them without even talking to them.

Consider HRV, a measure of the variation in time intervals between heartbeats. HRV is an indicator of autonomic nervous system (ANS) activity, reflecting the balance between the sympathetic (fight-or-flight) and parasympathetic (rest-and-digest) branches. Certain acoustic frequencies can stimulate the vagus nerve, which is a major component of the parasympathetic nervous system, leading to increased HRV and a state of relaxation. By monitoring HRV in real-time during acoustic therapy sessions, we can fine-tune the frequencies and intensities to achieve the desired physiological effect. Similarly, EEG can be used to assess brainwave activity in response to different sounds. Certain frequencies can promote alpha and theta brainwave activity, which are associated with relaxation and meditative states. EMG can be used to measure muscle tension, providing insights into how sound is affecting muscle relaxation and reducing pain.

The beauty of physiological signal analysis lies in its ability to personalize treatment. Every individual responds differently to sound, depending on their unique physiology, medical history, and current state of well-being. By monitoring their physiological responses in real-time, we can adjust the acoustic parameters to optimize the therapeutic effect. It is also important to use techniques that will have little to no effect on the patient. So, for example, if one wants to monitor heart sounds to establish if a patient is affected by a valve abnormality, the use of an electrocardiogram (ECG) could easily be replaced by a microphone. This has the beneficial effect of being a non-invasive procedure. However, there are downsides of this method, as microphones are more affected by signal noise.

The connection between acoustic stimulation and physiological responses is not always straightforward. Factors such as the individual's emotional state, the environment in which the therapy is delivered, and the specific characteristics of the sound itself (frequency, intensity, duration, waveform) can all influence the outcome. Therefore, a holistic approach is essential. I always tell my students to consider the individual as a whole, not just as a collection of physiological signals. It is important to be kind and sensitive, and this has a profound effect on patients.

Finally, the integration of acoustic therapy with other medical modalities holds immense promise. Combining acoustic therapy with conventional treatments, such as medication or physical therapy, may enhance their effectiveness and reduce the need for pharmaceutical interventions. For example, acoustic therapy could be used to manage pain and anxiety associated with cancer treatment, allowing patients to reduce their reliance on opioid medications. Or, it could be used to improve sleep quality and reduce stress in patients with cardiovascular disease, complementing traditional medical management.

In essence, acoustic therapy is an evolving field with the potential to revolutionize the way we approach medical treatment. By understanding the principles of sound and how it interacts with the human body, and by employing sophisticated physiological signal analysis techniques, we can unlock the therapeutic power of sound to promote healing and well-being. I believe that sound is the medicine of the future, and I am excited to be a part of this journey.

1.1 Acoustic Therapy: An Overview

Acoustic therapy holds a unique position within the realm of medical treatments, leveraging the power of sound to influence physiological processes. It's not just about playing soothing melodies; it encompasses a precise application of sound frequencies and intensities to target specific bodily functions and conditions. At its core, acoustic therapy hinges on the principle that sound vibrations can interact with cells, tissues, and organs, potentially altering their activity and promoting healing or symptom relief. The spectrum of applications is broad, ranging from pain management and stress reduction to potentially influencing cardiovascular and respiratory functions, as we'll explore further in subsequent chapters.

The lineage of acoustic therapy can be traced back through various historical practices, albeit often intertwined with less scientifically rigorous approaches. Ancient cultures recognized the potential of sound in healing rituals, utilizing chanting, drumming, and resonant instruments to influence mental and physical states. However, the modern understanding of acoustic therapy, as we consider it today, relies on a more evidence-based approach, grounded in the physics of sound and the biological responses it elicits. Significant advancements in sensor technology and signal processing have enabled us to precisely measure and analyze physiological responses to sound, paving the way for targeted acoustic interventions. This represents a departure from purely subjective experiences, moving towards a more quantifiable and reproducible therapeutic modality. The key differentiating factor between historical practices and modern acoustic therapy lies in the scientific understanding of how sound interacts with the human body at a cellular and systemic level, as well as the technological capabilities that allow for precise control and monitoring of acoustic stimuli.

Analyzing physiological signals is an indispensable component of modern acoustic therapy. It transforms what could be a nebulous and subjective treatment into a data-driven and personalized intervention. By monitoring bodily signals, such as heart rate variability, respiratory rate, brainwave activity, and muscle tension, we can gain insight into the patient's current physiological state and how they are responding to acoustic stimulation. For instance, measuring heart rate variability can reflect the balance between the sympathetic and parasympathetic nervous systems, providing a marker of stress or relaxation. Similarly, changes in brainwave patterns, as measured by electroencephalography (EEG), can indicate alterations in cognitive state and level of alertness. This real-time feedback allows therapists to fine-tune acoustic parameters, such as frequency, intensity, and duration, to achieve the desired therapeutic effect. The ultimate goal is to optimize the acoustic intervention based on the individual's unique physiological response, thereby maximizing its efficacy and minimizing potential side effects.

One of the major challenges lies in isolating the specific mechanisms by which sound exerts its therapeutic effects. While we know that sound vibrations can stimulate cells and trigger biochemical cascades, the precise pathways and molecular targets are not always fully understood. This lack of clarity makes it difficult to design and optimize acoustic therapy protocols for specific conditions. Consider, for example, the potential of acoustic stimulation to influence the inflammatory response. While some studies suggest that certain frequencies can reduce inflammation by modulating the activity of immune cells, the exact mechanisms and optimal parameters remain unclear. Further research is needed to elucidate these mechanisms and identify specific biomarkers that can predict and monitor the response to acoustic therapy. Another challenge is the variability in individual responses to sound. Factors such as age, gender, genetic predisposition, and pre-existing conditions can all influence how a person reacts to acoustic stimulation. This heterogeneity makes it difficult to develop standardized protocols that are effective for all patients. Personalized approaches, guided by physiological signal analysis, are essential to overcome this challenge and optimize therapeutic outcomes.

Looking ahead, several promising avenues of research and development hold the potential to revolutionize acoustic therapy. One exciting area is the use of focused ultrasound to deliver targeted acoustic energy to specific tissues and organs. This technique, known as histotripsy, can be used to non-invasively destroy tumors or break up blood clots. Another promising area is the development of closed-loop acoustic therapy systems that automatically adjust acoustic parameters based on real-time physiological feedback. Such systems could provide truly personalized and adaptive treatments, maximizing efficacy and minimizing the need for manual adjustments. Furthermore, the integration of acoustic therapy with other therapeutic modalities, such as medication, physical therapy, and psychotherapy, could lead to synergistic effects and improved outcomes. For example, combining acoustic stimulation with mindfulness practices may enhance the stress-reducing effects of both interventions. The future of acoustic therapy lies in its ability to harness the power of sound in a precise, targeted, and personalized manner, guided by a deep understanding of physiological responses and technological advancements.

1.2 Physiological Signal Analysis

Let's dive into the world of physiological signal analysis, a cornerstone of understanding and utilizing acoustic therapy devices. This isn't just about hearing sounds; it's about deciphering the language of the body, translated into electrical signals, and extracting meaningful insights for medical patient detection. It's where the artistry of medicine meets the precision of engineering.

At its core, physiological signal analysis involves capturing, processing, and interpreting signals that originate from the body. These signals reflect various physiological processes, offering valuable information about a person's health status. Think of it as eavesdropping on the body's internal conversations, where each murmur, each electrical pulse, tells a story. These signals come from many different sources, like the electrical rhythm of the heart, the mechanical movement of air in and out of the lungs, or the electrical activity of the brain. Each requires a different set of tools and methods to measure and to understand.

The first step in this process is signal acquisition. This involves using specialized sensors to detect and convert physiological activity into measurable electrical signals. For instance, electrocardiography (ECG) uses electrodes placed on the skin to capture the electrical activity of the heart. Similarly, electromyography (EMG) measures the electrical activity of muscles, while electroencephalography (EEG) records brainwave activity through electrodes on the scalp. Choosing the right sensor is critical; it must be sensitive enough to capture the subtle variations in the signal, yet robust enough to minimize interference from external noise. The quality of the acquired signal directly impacts the accuracy and reliability of subsequent analysis. It's like choosing the right microphone for a recording – a poor choice can ruin even the most talented performance.

Once the signals are acquired, the real work begins: signal processing. Raw physiological signals are often contaminated with noise and artifacts, making it difficult to extract meaningful information directly. Signal processing techniques are employed to clean up the signal, enhance relevant features, and remove unwanted noise. This may involve filtering techniques to remove high-frequency noise or baseline wander, artifact removal algorithms to eliminate distortions caused by movement or electrical interference, and signal averaging to improve the signal-to-noise ratio. Time-frequency analysis, such as wavelet transforms, can also be used to decompose the signal into its constituent frequency components, providing insights into the underlying physiological processes. For instance, analyzing heart rate variability (HRV) using time-frequency methods can reveal information about the autonomic nervous system's control of heart function. It's like a skilled audio engineer mastering a track, removing imperfections and bringing out the hidden nuances.

After signal processing, the next step is feature extraction. This involves identifying and quantifying relevant characteristics of the signal that can be used to discriminate between different physiological states or conditions. Features can be extracted from the time domain, frequency domain, or time-frequency domain, depending on the nature of the signal and the specific application. For example, in ECG analysis, features such as the R-R interval, QRS duration, and ST-segment elevation can be extracted to detect arrhythmias or ischemic heart disease. In EEG analysis, features such as the amplitude, frequency, and phase of different brainwave bands (e.g., alpha, beta, theta, delta) can be used to assess cognitive states, sleep stages, or neurological disorders. The extracted features serve as inputs to machine learning algorithms for classification and prediction. It's like distilling the essence of a complex situation into a few key data points that tell the story.

Now comes the application of machine learning. The extracted features are fed into machine learning algorithms to build predictive models that can automatically detect or classify different physiological states or conditions. Supervised learning algorithms, such as support vector machines (SVMs), decision trees, and neural networks, are commonly used for this purpose. These algorithms learn from labeled data, where each data point is associated with a known physiological state or condition. The trained model can then be used to classify new, unseen data and make predictions about a patient's health status. For instance, a machine learning model could be trained to detect sleep apnea from respiratory signals or to predict the risk of sudden cardiac death from ECG recordings. It's like training a detective to spot clues and solve a case, using past experience to make accurate deductions.

Of course, the work doesn't end there. The performance of the machine learning model must be rigorously evaluated to ensure its accuracy and reliability. This involves using metrics such as sensitivity, specificity, precision, and accuracy to assess the model's ability to correctly identify positive and negative cases. Cross-validation techniques are used to ensure that the model generalizes well to new data and does not overfit to the training data. It's like putting a new product through rigorous testing to ensure it meets the required standards.

Ethical considerations are crucial in physiological signal analysis, especially when dealing with sensitive patient data. Data privacy and security must be prioritized to protect patient confidentiality. Informed consent must be obtained from patients before collecting and analyzing their physiological signals. Additionally, the potential biases in the data and the algorithms must be addressed to ensure fairness and equity in the application of these technologies. It's like upholding the highest ethical standards in any profession, ensuring that technology is used responsibly and for the benefit of all.

And the future of physiological signal analysis is bright. Advances in sensor technology, signal processing algorithms, and machine learning techniques are continually expanding the possibilities for using physiological signals to improve medical patient detection. Wearable sensors and mobile health technologies are enabling continuous monitoring of physiological signals in real-world settings. The integration of physiological signal analysis with other data sources, such as genomic data and clinical records, is providing a more holistic view of patient health. Ultimately, physiological signal analysis holds the promise of transforming medical diagnostics and personalized medicine, enabling earlier detection of diseases, more effective treatments, and improved patient outcomes. It's an ever-evolving field, with the potential to revolutionize healthcare as we know it.

2 Chapter 2: Physiological Signals and Devices

Heart sound analysis plays a pivotal role in diagnosing a wide array of cardiovascular conditions. The intricate symphony of the heart, comprised of distinct sounds arising from valve closures and blood flow dynamics, provides invaluable insights into cardiac health. Think of the heart as an engine, and the sounds as the telltale signs of its operation; abnormal sounds often signal underlying mechanical or electrical malfunctions. Clinically, these sounds are traditionally assessed via auscultation using a stethoscope. However, advanced acoustic therapy devices offer a more precise and quantitative approach, enabling early detection and accurate monitoring of cardiac diseases. These devices typically employ sophisticated sensors and signal processing algorithms to capture, amplify, and analyze heart sounds, revealing subtle anomalies that may be missed during routine auscultation.

The primary components of a normal heart sound cycle are the first heart sound (S1) and the second heart sound (S2). S1, often described as a "lub," corresponds to the closure of the mitral and tricuspid valves, marking the beginning of systole. S2, characterized as a "dub," is associated with the closure of the aortic and pulmonic valves, signaling the onset of diastole. Pathological heart sounds, on the other hand, encompass a spectrum of abnormalities, including murmurs, extra sounds (S3 and S4), and clicks. Murmurs, which are turbulent blood flow sounds, can arise from valvular stenosis, regurgitation, or congenital heart defects. Third heart sounds (S3) can indicate ventricular overload or heart failure, while fourth heart sounds (S4) are often associated with atrial contraction against a stiff ventricle. Clicks are typically related to valve abnormalities. Analysis of these sounds, considering their timing, intensity, and location, yields valuable diagnostic clues.

Devices employed for heart sound analysis range from basic electronic stethoscopes to sophisticated phonocardiography systems. Electronic stethoscopes amplify heart sounds, enhancing audibility, particularly in noisy environments. Phonocardiography provides a graphical representation of heart sounds, enabling detailed visualization and analysis. Modern devices often incorporate advanced signal processing techniques such as noise reduction, filtering, and spectral analysis to enhance signal quality and extract relevant features. Sensors are also crucial, and should be considered, with microphones designed to detect heart sounds. Accelerometers can also be employed to measure chest wall vibrations related to cardiac activity, offering an alternative approach to acoustic signal acquisition. The most important feature to look for is the signal to noise ratio of the sensor.

Moving on to respiratory sound analysis, it's another crucial aspect of acoustic therapy, with implications for diagnosing and monitoring a variety of pulmonary conditions. The intricate sounds generated during breathing provide valuable information about the health of the respiratory system, much like listening to a musical instrument to assess its condition. Just as a skilled musician can identify subtle imperfections in tone, trained clinicians can discern pathological respiratory sounds, such as wheezes, crackles, and rhonchi, which are indicative of underlying respiratory disorders. These sounds arise from airflow obstruction, inflammation, or fluid accumulation in the lungs and airways.

Normal respiratory sounds include vesicular, bronchial, and bronchovesicular sounds. Vesicular sounds are soft, breezy sounds heard over the periphery of the lungs, representing normal airflow in the alveoli. Bronchial sounds, which are louder and harsher, are typically heard over the trachea and main bronchi. Bronchovesicular sounds are a blend of vesicular and bronchial sounds, heard near the sternum and between the scapulae. Abnormal respiratory sounds encompass a range of adventitious sounds, including wheezes, crackles (rales), rhonchi, and stridor. Wheezes are high-pitched, whistling sounds resulting from narrowed airways, often associated with asthma or COPD. Crackles, also known as rales, are discontinuous, popping sounds caused by the sudden opening of collapsed alveoli or the movement of fluid in the airways, commonly observed in pneumonia or pulmonary edema. Rhonchi are low-pitched, snoring-like sounds arising from airway obstruction due to mucus or secretions. Stridor is a high-pitched, harsh sound indicative of upper airway obstruction, requiring immediate medical attention.

Several devices can be used to assess respiratory sounds, from basic stethoscopes to computerized respiratory sound analysis systems. Stethoscopes remain the cornerstone of respiratory auscultation, providing a simple and cost-effective means of assessing lung sounds. Computerized systems offer more advanced capabilities, including sound recording, analysis, and visualization. These systems often incorporate sophisticated algorithms for noise reduction, signal enhancement, and automated detection of abnormal respiratory sounds. The sensors used in respiratory sound analysis typically consist of microphones placed on the chest wall to capture acoustic signals. Accelerometers can also be used to measure chest wall vibrations associated with breathing, providing complementary information. It's worth noting that the placement and sensitivity of sensors greatly influence the quality of the acquired signals.

Finally, brainwave monitoring, also known as electroencephalography (EEG), offers a unique window into the electrical activity of the brain, providing insights into brain function and neurological disorders. The brain, a complex network of interconnected neurons, generates electrical signals that can be detected by electrodes placed on the scalp. These signals, known as brainwaves, reflect the underlying neural activity and vary depending on the state of arousal, cognitive activity, and presence of neurological conditions. Analyzing these brainwaves allows us to detect seizures, sleep disorders, and other neurological conditions.

Brainwaves are categorized into different frequency bands, each associated with distinct mental states. Delta waves (0.5-4 Hz) are slow waves dominant during sleep. Theta waves (4-8 Hz) are observed during drowsiness, meditation, and early stages of sleep. Alpha waves (8-13 Hz) are associated with relaxed wakefulness and are prominent when the eyes are closed. Beta waves (13-30 Hz) are fast waves observed during active thinking, alertness, and concentration. Gamma waves (30-100 Hz) are the fastest waves and are associated with higher cognitive functions, such as perception and consciousness. Abnormal brainwave patterns can indicate neurological disorders. Seizures are characterized by abnormal, synchronized electrical activity in the brain. Sleep disorders, such as insomnia and sleep apnea, can alter the normal sleep EEG patterns. Brain tumors, head injuries, and other neurological conditions can also affect brainwave activity, leading to abnormal EEG patterns.

EEG devices consist of electrodes, amplifiers, and recording systems. Electrodes are placed on the scalp according to standardized layouts, such as the 10-20 system. These electrodes detect the electrical activity of the brain and transmit the signals to amplifiers. Amplifiers increase the amplitude of the brainwave signals, making them suitable for recording. Recording systems digitize and store the amplified brainwave signals for subsequent analysis. The types of electrodes vary widely and should be considered when designing a therapeutic device. Silver/silver chloride (Ag/AgCl) electrodes are commonly used due to their low noise characteristics and stable performance. Disposable electrodes are convenient for single-use applications, while reusable electrodes require proper cleaning and maintenance. Signal processing techniques are crucial for analyzing EEG data. Artifact removal is essential to eliminate noise and interference from sources such as muscle movements, eye blinks, and electrical interference. Filtering techniques can be used to isolate specific frequency bands of interest.

2.1 Heart Sound Analysis

Deciphering the language of the heart, beyond the simple lub-dub rhythm, is a complex yet rewarding endeavor. It's about uncovering subtle nuances within the sounds, those telltale signs that speak volumes about the heart's health. Think of it as becoming fluent in a secret cardiac dialect, where each variation in pitch, timing, and intensity holds diagnostic clues. To understand all of this we must get into the nature of the sounds, which have several qualities, the intensity (loudness), the pitch (frequency), timing, and duration.

To begin, let's grasp the fundamentals. The familiar "lub-dub" corresponds to the first heart sound (S1) and the second heart sound (S2), respectively.

S1 marks the closure of the mitral and tricuspid valves at the start of ventricular contraction (systole). S2 signifies the closure of the aortic and pulmonic valves at the beginning of ventricular relaxation (diastole). The interval between S1 and S2 represents systole, while the interval between S2 and the next S1 represents diastole. These sounds are normally audible with a stethoscope, providing a primary assessment of heart function. I remember my first time using a stethoscope – the pressure of potentially hearing something significant was immense.

However, the true artistry of heart sound analysis lies in detecting abnormal sounds, often referred to as murmurs or extra heart sounds. Murmurs are turbulent blood flow sounds, most commonly caused by valve defects (stenosis or regurgitation) or congenital heart abnormalities. Stenosis refers to the narrowing of a valve opening, restricting blood flow. Regurgitation, on the other hand, is the backflow of blood through a valve that doesn't close properly. Extra heart sounds, such as S3 and S4, can indicate specific cardiac conditions. An S3, also known as a ventricular gallop, often arises from rapid ventricular filling in conditions like heart failure. S4, or an atrial gallop, is typically associated with stiff or non-compliant ventricles, often seen in hypertension or hypertrophic cardiomyopathy.

The first step in mastering heart sound analysis involves honing your auscultation skills. This requires a quiet environment, a good-quality stethoscope (preferably with both bell and diaphragm), and a systematic approach. Begin by identifying S1 and S2. Pay close attention to their intensity, timing, and any splitting (when a single sound is heard as two distinct components). The diaphragm of the stethoscope is best for high-pitched sounds like S1 and S2, while the bell is more suited for low-pitched sounds like S3 and S4, or the rumble of mitral stenosis. Listen in all four cardiac auscultation areas: aortic (right second intercostal space), pulmonic (left second intercostal space), tricuspid (left lower sternal border), and mitral (apex of the heart). Focus on one area at a time, and then methodically move to the next.

Once you've identified the basic heart sounds, concentrate on detecting any murmurs. When you hear a murmur, determine its timing (systolic, diastolic, or continuous), shape (crescendo, decrescendo, plateau), intensity (graded from 1 to 6), pitch (high, medium, low), and location of maximal intensity. Systolic murmurs occur between S1 and S2, while diastolic murmurs occur between S2 and the next S1. Continuous murmurs extend throughout systole and diastole. The shape of a murmur describes its change in intensity over time. For example, a crescendo murmur gradually increases in intensity, while a decrescendo murmur gradually decreases.

The intensity of a murmur is graded on a scale of 1 to 6, with 1 being barely audible and 6 being audible without a stethoscope. However, it's important to remember that the intensity of a murmur doesn't necessarily correlate with the severity of the underlying condition. The pitch of a murmur is determined by the frequency of the sound. High-pitched murmurs are often associated with regurgitant lesions, while low-pitched murmurs are more commonly associated with stenotic lesions. The location of maximal intensity helps pinpoint the valve or structure responsible for the murmur. Mitral regurgitation, for example, is typically heard best at the apex of the heart, while aortic stenosis is loudest in the aortic area.

Distinguishing between different types of murmurs can be challenging, but there are several key features to consider. Aortic stenosis typically produces a systolic crescendo-decrescendo murmur that radiates to the neck. Mitral regurgitation, on the other hand, creates a systolic plateau murmur that radiates to the axilla. Mitral stenosis often presents with a diastolic rumble, best heard at the apex with the bell of the stethoscope, often accentuated by exercise. Aortic regurgitation typically causes a diastolic decrescendo murmur, best heard along the left sternal border.

Remember that heart sound analysis is not a standalone diagnostic tool. It should always be integrated with other clinical findings, such as the patient's history, physical examination, and other diagnostic tests like electrocardiography (ECG) and echocardiography. While auscultation is a crucial initial step, it often requires confirmation and further evaluation with more advanced imaging techniques.

As you gain experience, you'll begin to recognize subtle variations in heart sounds that can provide valuable clues to underlying cardiac conditions. You might hear a faint murmur in a young athlete, which turns out to be an innocent flow murmur. Or you might detect a subtle S4 in an elderly

patient, indicating underlying diastolic dysfunction. These are the moments when your honed skills can make a real difference in patient care.
Continuous learning is essential for mastering heart sound analysis. Attend conferences, read journals, and seek opportunities to practice with experienced clinicians. Listen to heart sounds whenever possible, and don't be afraid to ask questions. Use online resources to augment your training. There are several websites and mobile apps that offer recordings of various heart sounds and murmurs. Listening to these recordings can help you train your ear and improve your ability to distinguish between different types of heart sounds.
Finally, remember that patience and persistence are key. It takes time and dedication to develop the skills necessary to accurately analyze heart sounds. Don't get discouraged if you don't hear everything perfectly right away. The more you practice, the more confident and proficient you will become. Over time, you'll develop a "cardiac ear" that allows you to quickly and accurately assess heart sounds, contributing significantly to patient diagnosis and management. It is an amazing journey for those who dedicate themselves to it.

2.2 Respiratory Sound Analysis

Let's explore the intricate world of respiratory sound analysis. It's a crucial component in the broader field of acoustic therapy. It offers a non-invasive method to assess lung function and diagnose various respiratory ailments, which is vital for both initial diagnosis and continuous monitoring. The human respiratory system, with its complex network of airways, produces a rich tapestry of sounds during breathing. These sounds, often subtle and nuanced, can hold vital clues about the health and functionality of the lungs. Understanding how to accurately capture, analyze, and interpret these sounds is what makes respiratory sound analysis a powerful diagnostic tool.

Beginners in this area often find themselves overwhelmed by the sheer complexity of respiratory sounds, however, there are some key concepts to grasp that will help you navigate this field. Firstly, the distinction between normal and abnormal respiratory sounds is paramount. Normal breath sounds, like vesicular, bronchial, and bronchovesicular sounds, are produced by the movement of air through the airways. Abnormal sounds, on the other hand, such as wheezes, crackles (rales), rhonchi, and stridor, arise from various pathological conditions affecting the respiratory system. Wheezes, for example, are high-pitched whistling sounds caused by narrowed airways, often seen in asthma or COPD. Crackles, or rales, are discontinuous popping sounds that indicate fluid in the small airways or alveoli, common in conditions like pneumonia or pulmonary edema. Rhonchi are low-pitched, snoring-like sounds that suggest obstruction or secretions in the larger airways. Stridor is a high-pitched, harsh sound indicating upper airway obstruction. These sounds can vary in pitch, duration, intensity, and timing within the respiratory cycle, adding layers of complexity to their interpretation.

Beyond simply identifying the presence of abnormal sounds, it is important to understand the underlying physiological mechanisms responsible for their generation. Wheezes, for instance, are not simply a sound of narrowed airways, but rather the result of complex interactions between airflow, airway wall vibrations, and airway geometry. Crackles can be further subdivided into fine and coarse crackles, each indicating different pathological processes. Fine crackles are typically associated with interstitial lung diseases, while coarse crackles are more common in pneumonia. A deep understanding of these physiological correlations allows for more precise diagnostic inferences. Furthermore, consider the equipment used to capture these sounds. Traditional stethoscopes, while still widely used, have limitations in terms of amplifying and recording subtle sounds. Electronic stethoscopes, on the other hand, offer improved amplification, noise reduction, and recording capabilities, allowing for more detailed analysis. These tools allow for capturing of the sounds and storing them for comparison.

The next step involves analyzing the recorded respiratory sounds using signal processing techniques. This is where the raw acoustic data is transformed into a format that can be easily interpreted. Spectral analysis, for example, is a powerful tool for visualizing the frequency content of respiratory sounds. By plotting the amplitude of the sound at different frequencies, you can identify characteristic patterns associated with various respiratory conditions. Wheezes, for example, will typically appear as distinct peaks at specific frequencies. Time-frequency analysis, such as spectrograms or wavelet transforms, provides even more detailed information by showing how the frequency content of the sound changes over time. This is particularly useful for analyzing non-stationary sounds, such as crackles, which can vary in duration and timing. I have personally found that using a combination of spectral and time-frequency analysis provides the most comprehensive picture of the respiratory sound.

However, the analysis is not without its challenges. Noise contamination, both from environmental sources and internal body sounds (like heart sounds), can significantly degrade the quality of the recorded respiratory sounds. Noise reduction techniques, such as filtering and adaptive noise cancellation, are essential for minimizing the impact of noise on the analysis. Filtering involves selectively removing specific frequency ranges that are known to contain noise. Adaptive noise cancellation uses a reference signal to estimate and subtract the noise component from the respiratory sound. Another challenge is the variability in respiratory sounds among individuals. Factors such as age, body size, and breathing effort can influence the characteristics of the sounds. Therefore, it is crucial to establish baseline measurements and consider individual patient characteristics when interpreting respiratory sounds.

To overcome these limitations, researchers are increasingly turning to machine learning techniques for automated respiratory sound analysis. Machine learning algorithms can be trained on large datasets of respiratory sounds to automatically detect and classify various respiratory conditions. These algorithms can learn to identify subtle patterns and features that may be missed by human listeners. For example, a machine learning model could be trained to distinguish between different types of crackles or to detect early signs of respiratory distress. I believe machine learning holds immense potential for improving the accuracy and efficiency of respiratory sound analysis, but also consider the ethics involved when utilizing patient data.

In clinical practice, respiratory sound analysis can be used in a variety of settings. It can be used for screening patients for respiratory diseases, monitoring the effectiveness of treatment, and assessing the severity of respiratory conditions. For example, in asthma management, respiratory sound analysis can be used to detect early signs of airway narrowing and adjust medication dosages accordingly. In patients with pneumonia, it can be used to monitor the resolution of lung consolidation. As technology advances and devices become more portable and user-friendly, there is a growing trend towards remote patient monitoring using acoustic analysis devices. This enables healthcare providers to remotely monitor patients with chronic respiratory conditions, such as COPD, and intervene early if there are signs of deterioration. This can significantly improve patient outcomes and reduce the burden on healthcare systems.

Remember, respiratory sound analysis, although technically advanced, requires meticulous attention to detail. Accurate data acquisition and analysis are crucial for obtaining reliable results. Continuous training and education are essential to keep abreast of the latest advancements in the field. With practice and dedication, respiratory sound analysis can become a valuable tool in your diagnostic toolkit, ultimately contributing to better patient care and improved respiratory health. Understanding the complexities of acoustic profiles and the technology used to record them is essential to providing the best patient support.

2.3 Brainwave Monitoring

Brainwave monitoring, also known as electroencephalography (EEG), is a crucial technique for understanding brain activity and diagnosing neurological disorders, and is used as a diagnostic feature in acoustic therapy. The brain, a complex organ, generates electrical signals that reflect its various functions. These signals, when captured and analyzed, can reveal insights into cognitive processes, sleep patterns, and the presence of abnormalities. As we delve into the specifics of brainwave monitoring, it's important to recognize its role in complementing other physiological signal analyses, such as heart and respiratory sound analysis, offering a more holistic approach to patient assessment.

The fundamentals of EEG involve placing electrodes on the scalp to detect the minuscule electrical fluctuations produced by neuronal activity. These electrodes are strategically positioned according to standardized systems, like the 10-20 system, ensuring consistency and comparability across different recordings and studies. The raw EEG signals are then amplified and filtered to remove noise and artifacts, which can arise from muscle movements, eye blinks, or electrical interference. This pre-processing stage is critical for obtaining clean and interpretable data. Once the signals are cleaned, they are displayed as a series of waveforms, each corresponding to a specific brain region. These waveforms are characterized by their frequency, amplitude, and morphology, which are then interpreted by trained clinicians to identify patterns indicative of different brain states or pathologies. Understanding these fundamentals is the first step in appreciating the power of EEG as a diagnostic and monitoring tool.

Different brainwave frequencies correspond to distinct states of consciousness and activity. Delta waves (0.5-4 Hz) are the slowest and are prominent during deep sleep and in certain neurological conditions. Theta waves (4-8 Hz) are associated with drowsiness, meditation, and creative thinking. Alpha waves (8-12 Hz) are dominant during relaxed wakefulness with eyes closed. Beta waves (12-30 Hz) are present during active thinking, problem-solving, and alertness. Gamma waves (30-100 Hz) are the fastest and are linked to higher cognitive functions, such as perception and consciousness. The relative abundance and distribution of these brainwave frequencies can provide valuable information about an individual's mental state and neurological health. For instance, an excess of slow-wave activity (delta or theta) in an awake individual might suggest brain dysfunction or sleep deprivation. Conversely, a lack of alpha waves during relaxation could indicate anxiety or heightened alertness. By analyzing the spectral composition of EEG signals, clinicians can gain a deeper understanding of brain function.

The clinical applications of EEG are vast and varied, ranging from diagnosing epilepsy to monitoring sleep disorders and assessing brain injury. In epilepsy, EEG is used to identify seizure activity and classify different types of seizures, which is crucial for guiding treatment decisions. Characteristic EEG patterns, such as spike-and-wave discharges, are indicative of epileptic activity. EEG is also invaluable in sleep studies (polysomnography), where it helps to differentiate between sleep stages and identify sleep disorders like insomnia, sleep apnea, and narcolepsy. By monitoring brainwave activity, along with other physiological parameters, sleep specialists can accurately diagnose and manage these conditions. Furthermore, EEG plays a role in assessing the severity of brain injuries, such as traumatic brain injury (TBI) and stroke. It can help to identify areas of brain damage and monitor recovery over time. EEG can also be used to assess brain function in patients with cognitive disorders, such as dementia and Alzheimer's disease. While EEG is a powerful diagnostic tool, it's important to recognize its limitations. EEG primarily reflects cortical activity and may not be sensitive to deeper brain structures.

The integration of brainwave monitoring into acoustic therapy devices represents a promising avenue for enhancing patient care. By combining

EEG data with acoustic stimulation, it's possible to create personalized therapeutic interventions tailored to an individual's specific brainwave patterns. For example, in patients with anxiety or insomnia, acoustic therapy could be designed to promote alpha or theta wave activity, inducing a state of relaxation and improving sleep quality. In patients with cognitive impairment, acoustic stimulation could be used to enhance gamma wave activity, potentially improving cognitive function and memory. The challenge lies in developing sophisticated algorithms that can analyze EEG data in real-time and adjust acoustic parameters accordingly. This requires a deep understanding of the neural mechanisms underlying acoustic therapy and the development of adaptive control systems. Furthermore, it's essential to conduct rigorous clinical trials to validate the efficacy of these integrated approaches. Despite these challenges, the potential benefits of combining brainwave monitoring with acoustic therapy are immense. It could lead to more effective and personalized treatments for a wide range of neurological and psychiatric conditions.

Advancements in EEG technology are constantly expanding the possibilities for brainwave monitoring. Wearable EEG devices, such as headbands and ear-EEG, are becoming increasingly popular due to their convenience and ease of use. These devices allow for long-term monitoring in real-world settings, providing valuable insights into brain activity during daily life. High-density EEG systems, which use a large number of electrodes, offer improved spatial resolution, allowing for more precise localization of brain activity. Furthermore, new signal processing techniques are being developed to enhance the accuracy and reliability of EEG analysis. These include methods for artifact removal, source localization, and brain-computer interfaces. Source localization techniques aim to identify the specific brain regions that are generating the EEG signals, providing a more detailed understanding of brain function. Brain-computer interfaces (BCIs) use EEG signals to control external devices, such as computers or prosthetic limbs. BCIs have the potential to revolutionize the treatment of paralysis and other motor disorders. As technology continues to evolve, brainwave monitoring will undoubtedly play an increasingly important role in medical diagnostics and therapy.

Looking ahead, the future of brainwave monitoring is likely to be shaped by several key trends. One is the increasing integration of EEG with other neuroimaging techniques, such as fMRI and PET, to provide a more comprehensive picture of brain function. Another is the development of closed-loop systems that can automatically adjust therapeutic interventions based on real-time EEG feedback. These systems could be used to optimize drug delivery, electrical stimulation, or acoustic therapy. Furthermore, there is growing interest in using EEG for personalized medicine, tailoring treatments to an individual's unique brainwave profile. To achieve this, large databases of EEG data will be needed, along with sophisticated machine learning algorithms to identify patterns and predict treatment outcomes. Ethical considerations, such as data privacy and security, will also need to be addressed. As brainwave monitoring becomes more sophisticated and widely used, it's crucial to ensure that it is used responsibly and ethically. This requires ongoing dialogue between researchers, clinicians, policymakers, and the public.

3 Chapter 3: Acoustic Device Design Principles

Designing effective acoustic therapy devices requires a deep understanding of both the human physiology and the engineering principles behind sound capture and analysis. The goal is to create a system that can accurately detect and interpret subtle physiological changes indicative of a patient's condition, all while remaining non-invasive and user-friendly. This chapter will explore the essential elements of acoustic device design.

The core of any acoustic therapy device is its ability to transduce physical sound waves into electrical signals that can be processed and analyzed. Sensor selection is paramount. Contact microphones, also known as stethoscopes, directly contact the skin and convert mechanical vibrations into electrical signals. These are commonly used for capturing heart and lung sounds. Piezoelectric sensors, another viable option, generate a voltage when subjected to mechanical stress, offering high sensitivity and a wide frequency response. Accelerometers, which measure acceleration, can be used to detect vibrations associated with tremors or muscle activity.

MEMS (Micro-Electro-Mechanical Systems) microphones are increasingly popular due to their small size, low power consumption, and good sensitivity, making them suitable for wearable devices. Each sensor type has its own strengths and weaknesses. Contact microphones can be susceptible to environmental noise, while piezoelectric sensors may require impedance matching circuitry. Accelerometers can be sensitive to movement artifacts. MEMS microphones, while compact, may have limited dynamic range. Thus, careful consideration of the target application and environment is crucial when choosing a sensor. Factors such as sensitivity, frequency response, noise floor, size, power consumption, and cost must be weighed against the specific requirements of the acoustic therapy device.

Once the sound signal has been captured, it needs to be cleaned and prepared for analysis. Signal processing algorithms play a crucial role in enhancing the signal-to-noise ratio and extracting meaningful features. A common first step is filtering. Bandpass filters can isolate specific frequency ranges of interest, such as the heart sound frequencies (20-600 Hz) or lung sound frequencies (100-2000 Hz). Notch filters can remove specific noise frequencies, like the 60 Hz hum from power lines. Adaptive noise cancellation techniques can be employed to reduce background noise that varies over time. These algorithms use a reference signal to estimate and subtract the noise component from the desired signal. Another essential signal processing step is feature extraction. Time-domain features, such as signal amplitude, duration, and zero-crossing rate, can provide information about the characteristics of the sound signal. Frequency-domain features, obtained through techniques like Fast Fourier Transform (FFT), reveal the spectral content of the signal. Wavelet transforms provide a time-frequency representation, which can be useful for analyzing non-stationary signals. Cepstral analysis, commonly used in speech recognition, can extract features related to the vocal tract characteristics, which can be applied to lung sound analysis. Feature selection is crucial to reduce dimensionality and identify the most relevant features for patient state detection. Techniques like principal component analysis (PCA) or linear discriminant analysis (LDA) can be used to identify a smaller set of features that capture the most variance in the data.

Designing an acoustic therapy device is an iterative process. It requires careful consideration of sensor selection, signal processing algorithms, and user interface design. Here is my experience. During my time developing an acoustic device for asthma monitoring, it was crucial to balance sensitivity with practicality. Early iterations, which prioritizes high sensitivity sensors, were overly susceptible to ambient sounds – a cough from across the room, or a passing car would trigger it. Therefore, careful filtering and noise reduction techniques were essential. We ended up using a combination of adaptive noise cancellation and spectral subtraction, improving the signal-to-noise ratio considerably. Furthermore, the device needed to be comfortable and easy to use for patients of all ages. Bulky devices or overly complicated interfaces would lead to poor adherence.

We focused on creating a small, lightweight device that could be easily attached to the chest and controlled with a simple, intuitive interface. We conducted several rounds of user testing, which revealed areas for improvement in both the hardware and software. We found that providing real-time feedback on the quality of the signal was helpful for ensuring that patients were positioning the device correctly. I learned that the best device is useless if people don't use it.

Consider these crucial elements as you embark on the design of your acoustic therapy device. By carefully selecting the right sensor, implementing appropriate signal processing algorithms, and focusing on user-centered design, you can create a device that not only captures accurate physiological information but also provides a valuable tool for improving patient care.

Acoustic therapy devices are not just about electronics and algorithms. The physical housing plays a significant role in the overall performance and usability of the device. Ergonomics are paramount. The device should be comfortable to wear or hold for extended periods, especially if it is intended for continuous monitoring. The shape, size, and weight should be carefully considered to minimize discomfort and ensure proper placement. Materials selection is also important. Biocompatible materials should be used to avoid allergic reactions or skin irritation. The housing should be durable enough to withstand daily wear and tear, and it should be easy to clean and disinfect. Acoustic properties of the housing can also affect the performance of the device. The housing should be designed to minimize reflections and resonances that could distort the sound signal. The placement of the sensor within the housing is also critical. The sensor should be positioned to maximize its sensitivity to the target sound source while minimizing the pickup of unwanted noise. Shielding can be used to protect the sensor from electromagnetic interference. Connectivity is an important consideration. The device should be able to transmit data wirelessly to a smartphone, tablet, or computer for further processing and analysis. Bluetooth is a common choice for short-range wireless communication, while Wi-Fi can be used for longer-range communication. The device should also have a way to store data locally in case wireless connectivity is not available. Power management is also crucial. The device should be designed to minimize power consumption to extend battery life. Rechargeable batteries are preferable for devices that are intended for continuous use. The device should also have a way to indicate the battery level.

Finally, let's address the ethical considerations. It's not solely about capturing signals; it's about safeguarding the data. Ensuring data privacy is critical, employing robust encryption methods and adhering to healthcare regulations like HIPAA. Transparency is key. Patients should be fully informed about how their data is collected, used, and stored. Data security is paramount. Implement strict access controls and regular security audits to protect against unauthorized access and data breaches. Long-term storage requires careful planning. Ensure data integrity over time by using appropriate storage formats and backup strategies. Remember, building an acoustic therapy device is more than just engineering; it's about contributing to patient well-being in a responsible and ethical way.

3.1 Sensor Technology

Sensor technology stands as the crucial interface between the human body and the sophisticated signal processing systems that underpin acoustic therapy devices. Selection of the right sensor is paramount; it dictates the quality of the data acquired and, ultimately, the accuracy of diagnoses and the effectiveness of therapeutic interventions. Understanding the nuances of different sensor types and their suitability for specific applications is, therefore, not merely advantageous but absolutely essential for anyone involved in the design or application of these devices.

The realm of sensors applicable to acoustic therapy is vast, encompassing a diverse range of technologies, each with its own strengths and weaknesses. One of the most common types is the piezoelectric sensor, which converts mechanical pressure (such as vibrations from heart or lung sounds) into an electrical signal. These sensors are prized for their high sensitivity and relatively simple construction. However, they can be susceptible to noise from external sources and may require careful shielding to minimize interference. Another prevalent type is the microelectromechanical system (MEMS) microphone, which utilizes miniaturized mechanical structures to detect sound waves. MEMS microphones offer a good balance between size, cost, and performance, making them suitable for a wide range of applications. However, their sensitivity may be lower compared to piezoelectric sensors, particularly at lower frequencies. Furthermore, inductive sensors have emerged as a viable technology, converting the acoustic signal into electric signals by means of magnetic induction. This technology is known for its robustness and endurance, and its high level of reliability.

When choosing a sensor for a particular acoustic therapy application, several key factors must be considered. Sensitivity, or the ability to detect weak signals, is paramount. For example, detecting subtle heart murmurs requires sensors with exceptional sensitivity. Frequency response, which refers to the range of frequencies that the sensor can accurately detect, is another crucial consideration. Respiratory sounds, for instance, span a wide range of frequencies, necessitating sensors with a broad frequency response. Impedance matching, to ensure efficient signal transfer between the sensor and the signal processing circuitry, is also crucial. In addition, one must consider the environmental conditions in which the sensor will operate. For instance, humidity and temperature variations can affect sensor performance, and choosing sensors that are robust to these factors is essential for reliable operation. Moreover, power consumption will play a role, as the lower it is, the more durable the batteries will be, or a higher battery density can be used.

Beyond the sensor itself, proper signal conditioning is crucial for maximizing the quality of the acquired data. Signal conditioning involves a series of processes designed to amplify, filter, and otherwise modify the sensor signal to make it suitable for further processing. Amplification boosts the signal level, making it easier to detect and analyze. Filtering removes unwanted noise and interference, such as mains hum or muscle artifacts. Analog-to-digital conversion (ADC) converts the analog sensor signal into a digital format that can be processed by a computer. The resolution and sampling rate of the ADC are critical parameters that determine the accuracy and fidelity of the digitized signal. This also allows for wireless transmission, if necessary for remote monitoring.

Calibration is a vital step in ensuring the accuracy and reliability of sensor measurements. Calibration involves comparing the sensor output to a known standard and adjusting the sensor's response to match the standard as closely as possible. This process can be performed using specialized calibration equipment or by comparing the sensor's output to a reference signal. Regular calibration is essential to account for drift and other changes in sensor performance over time. Another aspect of this procedure is to find the sensors' bias, which can be corrected to produce more reliable data.

The placement of sensors is another critical factor that can significantly impact the quality of the acquired data. The ideal sensor location will depend on the specific physiological signal being measured. For example, heart sounds are typically best detected by placing the sensor on the chest wall over the heart. Respiratory sounds are best detected by placing the sensor on the back or chest, depending on the location of the lungs. Sensors should be securely attached to the body to minimize movement artifacts. Anatomical variations between patients and device location on the body can affect the amplitude and type of signal received.

Selecting the right sensor technology and implementing proper signal conditioning techniques are fundamental to building effective acoustic therapy devices. It is worth noting that new sensor materials and methodologies are in active development, such as carbon nanotubes, or using laser doppler vibrometry as a non contact sensor. Careful consideration of factors such as sensitivity, frequency response, environmental conditions, and sensor placement is essential for achieving accurate and reliable physiological signal measurements. The advancement of new sensor technologies also allows for smaller designs with higher data quality.

3.2 Signal Processing Algorithms

Signal processing algorithms are the unsung heroes behind every acoustic therapy device, the intricate codes transforming raw sensor data into clinically meaningful insights. Without them, the subtle sounds of the heart and lungs would remain a cacophony, indecipherable and useless for diagnosis or treatment. In this chapter, we'll delve into the specific algorithms that underpin these devices, focusing on those most relevant for extracting crucial physiological information.

Consider the challenge: a heart sound recording, potentially corrupted by ambient noise, muscle artifacts, or even the device's own internal electrical hum. Simply amplifying this raw signal won't help; it will only amplify the noise along with it. This is where signal processing comes in. Filtering is often the first step, employing algorithms designed to isolate the frequency bands of interest. For heart sounds, this typically involves bandpass filters targeting the range of 20-600 Hz, effectively attenuating both low-frequency rumble and high-frequency hiss. These filters can be implemented using various techniques, including Butterworth, Chebyshev, and Elliptic designs, each with its own trade-offs between sharpness of cutoff and ripple in the passband. The choice depends on the specific requirements of the application and the characteristics of the noise.

Beyond basic filtering, more sophisticated noise reduction techniques are often necessary. Adaptive filtering, for example, uses a reference signal – ideally correlated with the noise but not with the signal of interest – to estimate and subtract the noise component. Think of it like having a separate microphone picking up only the ambient noise, then subtracting that noise from the main recording. Another powerful approach is wavelet denoising, which decomposes the signal into different frequency scales and selectively removes noise components based on their statistical properties. This is particularly effective for removing transient noise spikes that might otherwise mask important features in the heart or lung sounds.

Once the signal is cleaned up, the next step is feature extraction. This involves identifying and quantifying specific characteristics of the signal that are relevant for diagnosis. For heart sounds, this might include the duration and amplitude of the first (S1) and second (S2) heart sounds, the presence and timing of any murmurs or extra sounds (S3, S4), and the intervals between these events. Algorithms for heart sound segmentation are crucial here, accurately identifying the boundaries of each sound component. These algorithms often rely on a combination of time-frequency analysis, envelope detection, and machine learning techniques. Similarly, for respiratory sounds, feature extraction might involve identifying the frequency and amplitude of wheezes, the presence and characteristics of crackles, and the overall airflow patterns.

But how do we reliably detect these features, especially when they are subtle or obscured by noise? Time-frequency analysis provides a powerful tool for visualizing the signal's spectral content over time. Techniques like the Short-Time Fourier Transform (STFT) and the Continuous Wavelet Transform (CWT) allow us to see how the frequency components of the signal change over time, revealing patterns that might not be apparent in the raw waveform. For example, a spectrogram of a respiratory sound might reveal the characteristic pattern of a wheeze, even if it's not immediately audible. I've found that these techniques are most effective when combined with expert knowledge of the underlying physiology. Knowing what to look for – the specific frequency range and temporal characteristics of a particular sound – can greatly improve the accuracy of feature extraction.

Further sophistication arises in the realm of advanced signal processing. Consider blind source separation (BSS) techniques. In scenarios where multiple physiological signals are captured simultaneously (e.g., heart and lung sounds recorded with a single sensor), BSS algorithms can attempt to separate these signals without prior knowledge of their characteristics. This can be particularly useful in noisy environments where cross-talk between different signals is a problem. Another important area is nonlinear signal processing. Many physiological signals exhibit nonlinear behavior, meaning that their response to stimuli is not proportional to the input. Traditional linear signal processing techniques may not be adequate for analyzing these signals, and nonlinear methods like fractal analysis and chaos theory may be required to extract meaningful information.

Thinking about real-world applications, the choice of algorithm also depends heavily on the computational constraints of the device. A portable acoustic therapy device that needs to operate on battery power will have different requirements than a high-performance workstation in a hospital setting. Real-time processing is often essential, requiring algorithms that are both accurate and computationally efficient. This often involves trade-offs between accuracy and complexity, and careful optimization of the code to minimize processing time and power consumption.

As a quick example, autocorrelation can be very useful for detecting periodicity in a signal, say, for measuring heart rate variability from a phonocardiogram. However, it can be computationally expensive. Therefore, simplified versions of autocorrelation might be used in real-time applications, such as using a sliding window approach to calculate autocorrelation only over short segments of the signal.

Finally, it's important to remember that signal processing algorithms are not a magic bullet. They are tools that must be used thoughtfully and in conjunction with other techniques, such as clinical expertise and medical imaging. The best results are achieved when the algorithms are carefully tailored to the specific application and validated on a representative dataset of patients. It's a process of continuous refinement, iterating between algorithm development, testing, and clinical validation. This iterative process, combined with a deep understanding of the underlying physiology, is what ultimately leads to the development of effective and reliable acoustic therapy devices.

Ultimately, the value of these algorithms lies in their ability to transform complex, often noisy, physiological signals into actionable information that can improve patient care. It's a challenging field, but one that offers enormous potential for advancing medical diagnostics and treatment.

4 Chapter 4: Patient Detection Methods

Alright, let's delve into patient detection methodologies.

Feature extraction forms the bedrock of any successful patient detection system relying on acoustic therapy devices. It's about isolating the most pertinent characteristics from the raw physiological signals we discussed earlier – heart sounds, respiratory patterns, and brainwaves, among others. Think of it as sifting through a mountain of data to find the gold nuggets. But what constitutes a "gold nugget" in this context? It depends heavily on the specific condition we're trying to detect. For instance, when analyzing heart sounds to detect a murmur, we're not just looking at the overall amplitude of the sound. We're meticulously examining its frequency content, duration, and timing relative to the heart cycle. We employ techniques like Fourier transforms to decompose the complex heart sound waveform into its constituent frequencies. The intensity of specific frequency bands, particularly those outside the normal range, can be indicative of valvular dysfunction or other structural abnormalities. Another valuable tool is wavelet analysis, which allows us to analyze signals at different scales or resolutions. This is particularly useful for detecting transient events like extra heart sounds or subtle changes in respiratory patterns that might be missed by traditional Fourier analysis. For respiratory sound analysis, we might focus on features like the presence and characteristics of wheezes, crackles, or stridor. We measure the duration, intensity, and frequency of these adventitious sounds and map their location within the respiratory cycle. These features can provide clues about the underlying pathology, such as airway obstruction, inflammation, or fluid accumulation. Finally, consider entropy, a measure of the signal's randomness or unpredictability. Increased entropy in heart rate variability, for example, can be an indicator of reduced cardiac function and increased risk of adverse events.

Having extracted these relevant features, the next step involves choosing the appropriate classification algorithm. This is where we teach the machine to differentiate between healthy and unhealthy states, or between different disease subtypes, based on the extracted features. No single classification algorithm is universally superior; the best choice depends on the characteristics of the data and the specific problem we're trying to solve. Let's examine some common algorithms. Logistic regression, despite its simplicity, can be surprisingly effective when dealing with linearly separable data. It models the probability of a patient belonging to a particular class (e.g., healthy or diseased) based on a linear combination of the input features. Support vector machines (SVMs) are particularly well-suited for high-dimensional data, which is often the case when dealing with physiological signals with many features. SVMs aim to find the optimal hyperplane that separates different classes of data points with the largest margin. Decision trees are intuitive and easy to interpret. They partition the feature space into a series of rectangular regions, each associated with a specific class label. Random forests, which are ensembles of multiple decision trees, often outperform single decision trees in terms of accuracy and robustness. Neural networks, inspired by the structure of the human brain, are capable of learning complex non-linear relationships between input features and output classes. Deep learning models, which are neural networks with multiple layers, have shown remarkable performance in a wide range of tasks, including image and speech recognition. When choosing a classification algorithm, it's essential to consider factors such as the size of the dataset, the dimensionality of the feature space, the presence of noise or outliers, and the desired level of interpretability. It's also important to experiment with different algorithms and evaluate their performance on a held-out validation set to select the best one for the task at hand.

So, you've extracted features, selected a classification algorithm, and trained your model. But how do you know if it's any good? This is where performance evaluation comes in. It's about rigorously assessing the accuracy, reliability, and generalizability of your patient detection system.

Several metrics are commonly used to evaluate classification performance. Accuracy, defined as the proportion of correctly classified instances, is a straightforward and intuitive measure. However, accuracy alone can be misleading, especially when dealing with imbalanced datasets where one class is much more prevalent than the other. Precision, also known as positive predictive value, measures the proportion of correctly identified positive cases out of all cases predicted as positive. Recall, also known as sensitivity or true positive rate, measures the proportion of correctly identified positive cases out of all actual positive cases. F1-score, the harmonic mean of precision and recall, provides a balanced measure of performance that takes into account both false positives and false negatives. The receiver operating characteristic (ROC) curve is a graphical representation of the trade-off between sensitivity and specificity at different classification thresholds. The area under the ROC curve (AUC) provides a single number that summarizes the overall performance of the classifier. In addition to these metrics, it's also important to consider the clinical relevance of the results. For example, a system with high sensitivity but low specificity might be acceptable for screening purposes, where it's more important to avoid missing any potential cases, even if it means generating a higher number of false positives. Conversely, a system with high specificity but low sensitivity might be preferred for diagnostic purposes, where it's crucial to minimize the number of false positives to avoid unnecessary interventions.

Ultimately, successful patient detection with acoustic therapy devices hinges on a careful combination of feature extraction, classification, and performance evaluation. By understanding the strengths and limitations of different techniques, you can build robust and reliable systems that improve patient outcomes and advance the field of medical diagnostics.

4.1 Feature Extraction Techniques

Feature extraction is the linchpin connecting raw physiological signals to actionable diagnostic insights. It's the process of transforming vast, complex datasets into a manageable set of representative characteristics, distilling the essence of the signal for effective analysis and classification. Think of it as separating the wheat from the chaff: identifying the key indicators that differentiate between healthy and diseased states. Without robust feature extraction, even the most sophisticated classification algorithms are rendered ineffective, struggling to discern meaningful patterns within the noise.

To begin our exploration, consider the raw complexity of heart sounds. A single heartbeat, captured by a stethoscope or electronic sensor, is a wave of acoustic information. It contains the familiar "lub-dub" – the S1 and S2 heart sounds – but also potentially subtle murmurs, clicks, and other anomalies. These anomalies, often masked by the dominant sounds, hold crucial diagnostic information about valve function, heart chamber size, and blood flow patterns. Successfully extracting these features requires more than just recording the sound; it requires sophisticated signal processing techniques. Time-domain analysis, frequency-domain analysis, and time-frequency analysis each offer unique perspectives on the acoustic data.

Time-domain features are often the first to be explored, owing to their simplicity and direct interpretability. These features operate directly on the amplitude of the signal as it changes over time. For instance, calculating the duration of the S1 and S2 sounds, the time interval between them (systole and diastole), and the overall heart rate can provide initial clues. The amplitude of different components of the signal can also be a powerful feature. However, time-domain features alone are often insufficient to capture the nuanced information present in complex heart sounds. Subtle murmurs, for example, might be easily missed if relying solely on basic time-domain metrics. The use of more complex techniques, such as dynamic time warping, can be used to classify heart sounds, without the requirement of knowing where in time the S1 or S2 signals occur.

Frequency-domain analysis, on the other hand, transforms the signal from the time domain to the frequency domain, revealing the constituent frequencies and their respective amplitudes. This is typically achieved using the Fast Fourier Transform (FFT). Specific frequency bands may be more prominent in certain cardiac conditions. For example, a heart murmur caused by valve regurgitation might exhibit increased energy in a particular frequency range. By calculating the power spectral density

(PSD), we can quantify the energy distribution across different frequencies, providing a richer set of features for classification. Frequency domain analysis can be effective in differentiating between healthy and abnormal heart sounds, even when time-domain features appear similar. However, frequency-domain analysis sacrifices temporal resolution; it provides information about the frequencies present in the signal but not when they occur.

To overcome the limitations of both time and frequency domain analysis, time-frequency analysis provides a hybrid approach. Techniques like the Short-Time Fourier Transform (STFT) and wavelet transforms decompose the signal into both time and frequency components. This allows us to track how the frequency content of the signal changes over time, revealing transient events and non-stationary characteristics. Wavelet transforms, in particular, are well-suited for analyzing non-stationary signals like heart sounds, as they offer multi-resolution analysis, providing both detailed and coarse-grained representations of the signal. Features extracted from time-frequency representations can include spectrograms, scalograms, and wavelet coefficients, which capture both the temporal and spectral dynamics of the signal. The use of higher order spectra analysis can improve the signal to noise ratio, by amplifying the signal, which allows better accuracy.

Turning to respiratory sounds, we encounter another set of challenges. Lung sounds, like heart sounds, are complex and often noisy, influenced by factors such as breathing rate, body position, and ambient noise. Normal breath sounds, such as vesicular and bronchial sounds, provide a baseline for comparison. Abnormal sounds, like wheezes, crackles, and stridor, indicate underlying respiratory conditions. Wheezes, characterized by their high-pitched, whistling sound, are often associated with airway obstruction, such as in asthma or COPD. Crackles, on the other hand, are short, discontinuous sounds indicative of fluid accumulation in the lungs, as seen in pneumonia or pulmonary edema. Stridor, a harsh, high-pitched sound, typically indicates upper airway obstruction.

Similar to heart sounds, feature extraction from respiratory sounds involves time-domain, frequency-domain, and time-frequency analysis. Time-domain features include the duration of inspiration and expiration, the ratio of inspiratory to expiratory time, and the amplitude of the sound during different phases of the respiratory cycle. Frequency-domain features, derived from the FFT, can reveal the dominant frequencies associated with different respiratory sounds. For example, wheezes typically exhibit higher frequencies compared to normal breath sounds. Time-frequency analysis, using techniques like spectrograms and wavelet transforms, can capture the temporal evolution of wheezes and crackles, providing valuable diagnostic information. However, the analysis of respiratory sounds are not as accurate as heart sounds, due to having a lower signal to noise ratio, and are more prone to distortion.

Beyond these traditional methods, advanced signal processing techniques are constantly being developed to improve feature extraction from physiological signals. These include techniques like Empirical Mode Decomposition (EMD), which decomposes a signal into a set of intrinsic mode functions (IMFs), each representing a different scale of variation in the signal. Features extracted from the IMFs can provide insights into the underlying physiological processes. Another promising area is the use of entropy-based measures, such as sample entropy and multiscale entropy, to quantify the complexity and irregularity of physiological signals. Changes in entropy can be indicative of disease states.

Finally, feature selection is a crucial step in the process. Once a large set of features has been extracted, it is important to select the most relevant and informative features for classification. Feature selection techniques aim to reduce the dimensionality of the feature space, improving the performance and interpretability of the classification model. Techniques like principal component analysis (PCA), linear discriminant analysis (LDA), and sequential feature selection (SFS) can be used to identify the most discriminative features. The choice of feature selection technique depends on the specific dataset and classification algorithm. Remember that the ultimate goal is to extract features that capture the essential characteristics of the physiological signals, enabling accurate and reliable patient detection. This involves a thoughtful combination of signal processing techniques, domain knowledge, and careful evaluation of the extracted features.

4.2 Classification Algorithms

Now, let's classify. Using features extracted from heart sounds, respiratory patterns, or even subtle variations in brainwave activity, we can train algorithms to automatically detect specific medical conditions. The power of these algorithms lies in their ability to sift through vast amounts of data, identifying patterns that might be imperceptible to the human ear or eye. This chapter serves as a deep dive into several classification techniques applicable to the analysis of acoustic therapy data.

To effectively categorize physiological signals, it's vital to understand the spectrum of classification algorithms available. Let's begin with a look at Support Vector Machines (SVMs). SVMs are particularly useful when dealing with high-dimensional data, which is a common occurrence in physiological signal analysis. Imagine plotting each data point, each representing a set of features extracted from a heart sound, on a multi-dimensional graph. The SVM algorithm then aims to find the optimal hyperplane that separates the different classes, for instance, healthy hearts from those with murmurs. The beauty of SVMs is their ability to handle non-linear data through the use of kernel functions, which map the input data into a higher-dimensional space where a linear separation becomes possible. SVMs work well with structured and unstructured data.

Next, we'll explore decision trees. Think of a decision tree like a flowchart, where each node represents a question about the data. For example, "Is the S1 heart sound amplitude above a certain threshold?". Based on the answer, the algorithm branches off to another node, asking another question, until finally, it reaches a leaf node, which represents the predicted class. Decision trees are easy to interpret and visualize, making them a valuable tool for understanding the decision-making process behind the classification. However, they can be prone to overfitting if the tree becomes too complex. The C4.5 and Classification and Regression Tree are popular decision tree algorithms.

Then there are neural networks. Inspired by the structure of the human brain, neural networks consist of interconnected nodes, or neurons, organized in layers. The input layer receives the raw data, which is then processed through multiple hidden layers, finally reaching the output layer, which provides the classification result. Neural networks are incredibly powerful and can learn complex patterns from data, but they also require a significant amount of training data to achieve optimal performance. Further, interpreting the decision-making process of a neural network can be challenging.

Moving forward, consider k-Nearest Neighbors (k-NN). This algorithm is remarkably simple yet effective. To classify a new data point, the k-NN algorithm finds the 'k' nearest data points in the training set and assigns the new data point to the class that is most common among those 'k' neighbors. The choice of 'k' is crucial; a small value can make the algorithm sensitive to noise, while a large value can smooth out the decision boundaries. This method is easy to implement, but choosing an optimal value for 'k' is paramount.

Naïve Bayes classifiers are another option to consider. This algorithm is based on Bayes' theorem, which calculates the probability of a data point belonging to a particular class, given its features. The "naïve" part comes from the assumption that all features are independent of each other, which is often not true in reality. Despite this simplification, Naïve Bayes classifiers can perform surprisingly well, especially in high-dimensional data. They are computationally efficient, making them suitable for real-time applications.

Selecting the right classification algorithm isn't a shot in the dark. It is all about understanding the characteristics of your data and the specific requirements of your application. Consider the size of your dataset, the complexity of the relationships between features, and the desired level of accuracy and interpretability. Sometimes, you might even need to experiment with different algorithms and compare their performance using appropriate evaluation metrics. The final choice depends on multiple factors.

Once you've chosen an algorithm, you will want to assess it's performance. Accuracy, precision, recall, and F1-score are your best friends. Accuracy measures the overall correctness of the classifier, while precision indicates the proportion of correctly classified positive instances. Recall, on the other hand, measures the proportion of actual positive instances that were correctly classified. The F1-score is the harmonic mean of precision and recall, providing a balanced measure of the classifier's performance. Receiver Operating Characteristic (ROC) curves are incredibly useful for visualizing the performance of a classifier at different threshold settings. The area under the ROC curve (AUC) provides a single number that summarizes the overall performance of the classifier. A higher AUC indicates better performance.

Cross-validation is an indispensable technique for evaluating the generalization performance of a classifier. It involves splitting the data into multiple folds, training the classifier on some folds, and testing it on the remaining folds. This process is repeated multiple times, with different folds used for training and testing each time. The results are then averaged to obtain a more robust estimate of the classifier's performance. Proper use of cross validation techniques can prevent overfitting and lead to more accurate future results.

Ultimately, classification algorithms are more than just mathematical formulas; they're powerful tools that can help us unlock valuable insights from physiological data. By carefully selecting and evaluating these algorithms, we can pave the way for earlier and more accurate diagnoses, leading to improved patient outcomes. These types of algorithms could be useful in preventative care. As such, they deserve to be implemented where they can.

4.3 Performance Evaluation

Evaluating the performance of patient detection methods is crucial. It's how we determine whether our fancy algorithms and meticulously designed acoustic devices are actually doing their job: accurately identifying patients with specific conditions based on their physiological signals. This isn't just about getting a "yes" or "no" answer; it's about understanding how reliable and trustworthy our system is in a clinical setting. We need to know its strengths and weaknesses, and how it compares to existing diagnostic tools. This informs further development, refinement, and ultimately, its responsible integration into healthcare.

One of the first and most fundamental metrics is accuracy. It's a simple concept: out of all the patients, how many did the system correctly classify? It sounds straightforward, but it's essential to consider the context. If we're screening for a rare disease, even a high accuracy might mask a poor performance in identifying the few affected individuals. Accuracy is calculated by dividing the number of correctly classified cases by the total number of cases evaluated. I find that plotting accuracy against different patient demographics can quickly reveal if the algorithm performs better in one subset of the population.

However, accuracy alone is rarely sufficient. Consider a scenario where 95% of the patients are healthy, and our system simply classifies everyone as healthy. It would achieve 95% accuracy, which sounds impressive, but it would completely fail to identify the patients who actually need treatment. This is where precision and recall come in, as I have seen repeatedly with several attempts to build medical tools using physiological measurements. Precision answers the question: of all the patients identified as having the condition, how many actually have it? This is important because a low precision means there will be many false positives, leading to unnecessary further testing and patient anxiety. Recall, on the other hand, asks: of all the patients who actually have the condition, how many were correctly identified by the system? A low recall means there will be many false negatives, potentially delaying treatment for patients who need it. Mathematically, precision is the number of true positives divided by the sum of true positives and false positives. Recall is the number of true positives divided by the sum of true positives and false negatives.

F1-score provides a balanced view. It's the harmonic mean of precision and recall, offering a single metric that considers both false positives and false negatives. This is particularly useful when there's an uneven distribution of the condition in the population, or when the cost of false positives and false negatives are different. A high F1-score indicates that the system has both good precision and good recall. The F1-score is calculated as 2 \emph{ (Precision } Recall) / (Precision + Recall). It is valuable to consider the F1-score in relation to both the precision and recall values. If there is a need to maximize one of those values, the F1-score can inform the potential downside to making that trade-off.

Receiver Operating Characteristic (ROC) curves and Area Under the Curve (AUC) are vital for understanding the system's ability to discriminate between patients with and without the condition. The ROC curve plots the true positive rate (sensitivity) against the false positive rate (1 \item specificity) at various threshold settings. AUC represents the area under the ROC curve and provides a single number that summarizes the overall performance. An AUC of 1 indicates perfect discrimination, while an AUC of 0.5 indicates performance no better than random chance. What I find particularly helpful about ROC curves is their visual representation of the trade-off between sensitivity and specificity. Depending on the clinical context, one might be prioritized over the other, and the ROC curve allows for informed decision-making regarding the optimal threshold.

Specificity and Sensitivity are other cornerstones of performance evaluation. Sensitivity, also known as the true positive rate, measures the proportion of actual positives that are correctly identified as such. Specificity, or the true negative rate, measures the proportion of actual negatives that are correctly identified as such. A highly sensitive test is good at detecting the condition when it's present, while a highly specific test is good at ruling out the condition when it's absent. I remember one project where we were developing a system for early detection of pneumonia. In that case, we prioritized sensitivity because missing a case of pneumonia could have severe consequences.

Beyond these basic metrics, there are other considerations. Calibration refers to how well the predicted probabilities from the system align with the actual probabilities. A well-calibrated system will, for example, predict a 70% probability of having the condition for a group of patients, and roughly 70% of those patients will actually have the condition. Calibration is particularly important when using the system's output for decision-making, as it allows clinicians to trust the predicted probabilities. I often use calibration curves to visually assess the calibration of a system.

Computational cost is also essential. How much time and resources does the system require to process the data and make a prediction? This is especially important in resource-constrained settings or when dealing with large datasets. A highly accurate system that takes hours to process a single patient's data might not be practical for real-world use.

Finally, clinical relevance is paramount. Ultimately, the goal is to improve patient outcomes. Does the system actually lead to earlier diagnosis, more effective treatment, or improved quality of life? Clinical trials and real-world implementations are necessary to demonstrate the clinical value of the system. One way to tackle this is by comparing the performance of the acoustic-based patient detection to the standard detection methods, in a way that does not sacrifice patient safety or outcomes.

Ultimately, evaluating performance is a multi-faceted process that requires a careful consideration of various metrics and their clinical implications. There is no single "best" metric, and the appropriate choice depends on the specific context and goals. The most effective approach is to combine multiple metrics and to interpret them in light of the clinical scenario. By rigorously evaluating performance, we can ensure that these advanced techniques are safe, effective, and beneficial for patients.

5 Chapter 5: Applications in Medical Diagnostics

The power of acoustic devices truly shines when we consider their diagnostic capabilities. Think about the stethoscope, a rudimentary acoustic device, yet still fundamental in preliminary medical assessments. Now imagine a more sophisticated version of this, enhanced by advanced signal processing and machine learning algorithms, capable of identifying subtle nuances in physiological sounds indicative of underlying pathology. That's the promise of modern acoustic therapy devices for medical diagnostics. We can push the boundaries of what's possible in patient care with such instruments.

Cardiovascular disease is a leading cause of mortality worldwide, and early detection is paramount for effective intervention. Traditional diagnostic methods, such as electrocardiograms (ECGs) and echocardiograms, are valuable but often require specialized equipment and trained personnel. Acoustic analysis offers a potentially more accessible and cost-effective approach. Imagine a device that can passively monitor heart sounds and detect early signs of valve stenosis or regurgitation. These conditions create turbulent blood flow, which generates characteristic murmurs. Advanced acoustic analysis techniques can extract features from these murmurs, such as their timing, duration, and frequency content, and use these features to train classification algorithms. These algorithms can then differentiate between normal and abnormal heart sounds with high accuracy. Let's dive into specific conditions and how acoustic analysis can aid in their detection. For instance, in aortic stenosis, the narrowing of the aortic valve creates a harsh, systolic murmur. The intensity of this murmur often correlates with the severity of the stenosis. Acoustic devices can quantify the murmur's intensity and potentially track its progression over time, providing valuable information for clinical decision-making. Similarly, mitral regurgitation, where blood leaks back into the left atrium during ventricular contraction, produces a characteristic holosystolic murmur. Acoustic analysis can identify the specific frequency components associated with this murmur and differentiate it from other types of heart sounds. Furthermore, heart failure, a complex syndrome characterized by the heart's inability to pump sufficient blood to meet the body's needs, often manifests with subtle changes in heart sounds. Acoustic analysis can detect these changes, such as the presence of an S3 gallop, an abnormal sound associated with rapid ventricular filling. By integrating acoustic data with other clinical information, physicians can gain a more comprehensive understanding of the patient's condition and tailor treatment accordingly.

Respiratory diseases, like cardiovascular ailments, also pose a significant public health challenge. Conditions such as asthma, chronic obstructive pulmonary disease (COPD), and pneumonia affect millions of people worldwide, and accurate diagnosis is crucial for effective management. Auscultation, the traditional method of listening to lung sounds with a stethoscope, remains an essential part of the respiratory examination. However, it is subjective and relies heavily on the clinician's experience. Acoustic analysis provides a more objective and quantitative approach to respiratory disease diagnosis. Consider asthma, a chronic inflammatory disorder of the airways characterized by airflow obstruction. During an asthma exacerbation, the airways narrow, leading to wheezing sounds caused by turbulent airflow. Acoustic devices can detect and quantify these wheezing sounds, providing an objective measure of airflow limitation. By analyzing the frequency, intensity, and duration of wheezes, clinicians can assess the severity of the asthma exacerbation and monitor the patient's response to treatment. Similarly, COPD, a progressive lung disease characterized by chronic bronchitis and emphysema, often manifests with crackles, also known as rales, which are discontinuous popping sounds caused by the opening and closing of small airways. Acoustic analysis can differentiate between different types of crackles and correlate them with the severity of the COPD. Furthermore, pneumonia, an infection of the lungs, can cause consolidation, where the airspaces become filled with fluid. This consolidation alters the transmission of sound through the lungs, leading to changes in breath sounds. Acoustic devices can detect these changes, such as bronchial breath sounds or decreased breath sounds, which can aid in the diagnosis of pneumonia. Acoustic analysis can also differentiate between different types of pneumonia, such as bacterial pneumonia, which often presents with localized consolidation, and viral pneumonia, which tends to be more diffuse. Imagine a scenario where a patient presents to the emergency department with shortness of breath. A quick acoustic assessment can help differentiate between asthma, COPD, and pneumonia, guiding the initial management strategy. Moreover, acoustic devices can be used for continuous monitoring of respiratory function, particularly in patients with chronic respiratory diseases. This allows for early detection of exacerbations and timely intervention, potentially preventing hospitalizations.

5.1 Cardiovascular Disease Detection

Cardiovascular disease (CVD) remains a leading cause of mortality worldwide. Early and accurate detection is paramount for effective management and improved patient outcomes. Acoustic therapy devices, leveraging the analysis of heart sounds, offer a non-invasive and potentially cost-effective approach to CVD screening and diagnosis. This chapter will delve into the specific applications of acoustic analysis for detecting various cardiovascular conditions, focusing on how these devices can contribute to improved patient care.

Analyzing heart sounds goes beyond simply listening with a stethoscope. Acoustic therapy devices, equipped with sensitive sensors and sophisticated algorithms, can capture and process heart sounds with far greater precision. This allows for the identification of subtle anomalies that might be missed during a traditional auscultation exam. One of the critical steps involves accurate sensor placement. While the specific location may vary depending on the device and the clinical context, standard auscultation points, such as the aortic, pulmonic, tricuspid, and mitral areas, serve as valuable starting points. Optimal sensor placement is crucial for capturing the full spectrum of heart sounds and maximizing diagnostic accuracy.

A key application lies in the detection of heart murmurs, abnormal sounds caused by turbulent blood flow within the heart. These murmurs can be indicative of valve stenosis (narrowing), valve regurgitation (leakage), or congenital heart defects. Acoustic analysis can differentiate between different types of murmurs based on their timing (systolic vs. diastolic), shape (crescendo-decrescendo, plateau), and location of maximal intensity. For instance, the distinct high-pitched, blowing diastolic murmur of aortic regurgitation can be distinguished from the low-pitched, rumbling diastolic murmur of mitral stenosis through meticulous acoustic analysis. Moreover, the intensity of the murmur, quantified through signal processing techniques, can provide an indication of the severity of the valve dysfunction. Feature extraction algorithms are essential for differentiating between normal and abnormal heart sounds. These algorithms often focus on identifying key characteristics of the heart sounds, such as the duration and amplitude of the S1 and S2 sounds, the presence of extra heart sounds (S3 and S4), and the characteristics of any murmurs present. Time-frequency analysis techniques, such as spectrograms and wavelet transforms, can provide detailed insights into the frequency content of heart sounds, enabling the detection of subtle abnormalities that might not be apparent in the time domain.

Beyond valve abnormalities, acoustic analysis can contribute to the detection of heart failure, a condition in which the heart is unable to pump enough blood to meet the body's needs. In heart failure, the presence of an S3 heart sound, also known as a ventricular gallop, is a strong indicator of increased ventricular filling pressure. Acoustic devices can reliably detect and quantify the S3 sound, providing valuable diagnostic information. Furthermore, changes in the timing and intensity of the normal heart sounds (S1 and S2) can also be indicative of heart failure. For example, a weakened S1 sound may suggest impaired ventricular contractility, a common feature of heart failure. Acoustic analysis, combined with other clinical data, can aid in the early diagnosis and management of heart failure.

Another area where acoustic therapy devices show promise is in the detection of coronary artery disease (CAD), a condition characterized by the buildup of plaque in the arteries that supply blood to the heart. While acoustic analysis cannot directly visualize the coronary arteries, it can detect subtle changes in heart sounds that may be associated with CAD. For instance, myocardial ischemia (reduced blood flow to the heart muscle) can alter the timing and intensity of heart sounds. Furthermore, the presence of an S4 heart sound, also known as an atrial gallop, may indicate reduced ventricular compliance, which can be a sign of CAD. While acoustic analysis is not a replacement for traditional diagnostic methods such as angiography, it can serve as a valuable screening tool for identifying individuals at risk for CAD.

The effectiveness of these devices hinges on accurate signal processing and the implementation of robust classification algorithms. Data from sensors needs careful filtering to remove noise and artifacts, such as those caused by movement or external sounds. Then, machine learning algorithms like support vector machines (SVMs) or neural networks can be trained to classify heart sounds as normal or abnormal, based on the extracted features. It's very important to validate these algorithms with large datasets of heart sounds from both healthy individuals and patients with various cardiovascular conditions, so we can ensure their accuracy and reliability. The device performance is evaluated using metrics such as sensitivity (the ability to correctly identify individuals with the disease), specificity (the ability to correctly identify individuals without the disease), and overall accuracy. These are used to evaluate the effectiveness of the algorithms in different contexts.

For the adoption of acoustic therapy devices, is very important to do clinical validation. Rigorous clinical trials are essential to demonstrate the accuracy and reliability of these devices in real-world settings. These trials should involve diverse patient populations and compare the performance of acoustic analysis to established diagnostic methods such as echocardiography and cardiac catheterization. Furthermore, it's important to evaluate the cost-effectiveness of these devices and to assess their impact on patient outcomes. If clinical trials demonstrate that acoustic therapy devices can improve the early detection of cardiovascular disease and lead to better patient management, their adoption in clinical practice will be significantly accelerated.

The potential advantages of acoustic therapy devices in CVD detection extend beyond diagnostic accuracy. They offer a non-invasive and potentially cost-effective alternative to traditional methods. The devices are easy to operate and suitable for use in a variety of settings, including primary care clinics, emergency departments, and even at home. This accessibility could lead to earlier detection of CVD, particularly in underserved populations who may not have easy access to specialized cardiac care. Moreover, the use of these devices can reduce the need for more invasive and costly diagnostic procedures, leading to significant cost savings for the healthcare system.

5.2 Respiratory Disease Diagnosis

Pinpointing respiratory illnesses using acoustic analysis is like listening to the intricate sounds of an orchestra; each instrument (or in this case, each part of the respiratory system) contributes to a complex symphony. When one instrument is out of tune, the entire piece suffers. Similarly, abnormal respiratory sounds can indicate specific pathologies, provided we know how to listen. For beginners, the first hurdle is distinguishing between normal and abnormal sounds, a skill that requires practice and a solid understanding of respiratory physiology.

The human respiratory system is a finely tuned mechanism, and its sounds reflect its health. Normal breathing produces relatively quiet, consistent sounds. These sounds are typically categorized as vesicular, bronchial, or bronchovesicular, depending on where they are auscultated on the chest. Vesicular sounds are soft, breezy sounds heard over the lung periphery, representing airflow through the small bronchioles and alveoli. Bronchial sounds are louder and harsher, typically heard over the trachea. Bronchovesicular sounds are a mix of both, usually heard near the major airways. Deviations from these normal sounds are where the diagnostic clues lie.

Understanding the common abnormal respiratory sounds is paramount. Wheezes, for instance, are high-pitched whistling sounds, often indicative of narrowed airways, as seen in asthma or COPD. They result from air being forced through constricted passages. Crackles, also known as rales, are discontinuous, popping sounds that can be either fine or coarse. Fine crackles are often associated with interstitial lung diseases or early-stage pneumonia, while coarse crackles suggest fluid accumulation in the larger airways, common in conditions like bronchitis or pulmonary edema. Stridor, a high-pitched, harsh sound heard primarily during inspiration, signals upper airway obstruction, a potentially life-threatening condition requiring immediate attention. Pleural rubs, which are grating or squeaking sounds, arise from inflamed pleural surfaces rubbing against each other during breathing, often indicating pleurisy.

To translate these sounds into a diagnosis, we must move beyond simple identification and consider the context. The location, timing within the respiratory cycle, and associated symptoms are all crucial pieces of the puzzle. For example, wheezing that is primarily expiratory suggests airway obstruction, such as in asthma, while inspiratory wheezing could point to a different underlying cause, such as vocal cord dysfunction. Crackles heard at the lung bases are more likely to be associated with heart failure or pulmonary fibrosis, whereas crackles heard throughout the lungs might indicate pneumonia. Paying attention to the patient's history, including their smoking habits, exposure to environmental pollutants, and past medical conditions, provides additional context for interpreting the sounds.

Now, let's talk about how acoustic therapy devices come into play. These devices are not just fancy stethoscopes; they employ advanced sensor technology and signal processing algorithms to capture and analyze respiratory sounds with greater precision. Think of it as enhancing our auditory capabilities with sophisticated technology. The sensors in these devices are designed to be highly sensitive to subtle variations in sound, picking up nuances that might be missed with a traditional stethoscope. The signal processing algorithms then filter out background noise, amplify relevant sounds, and even visualize the sound waves for further analysis.

Feature extraction techniques are crucial in these devices. These techniques involve identifying and quantifying specific characteristics of the respiratory sounds, such as the frequency, amplitude, and duration of wheezes or crackles. These features can then be used to train classification algorithms to distinguish between different respiratory diseases. For instance, an algorithm might be trained to differentiate between asthma and COPD based on the characteristics of the wheezes, or between pneumonia and pulmonary edema based on the type and distribution of crackles.

However, a word of caution: these devices are tools, not replacements for clinical judgment. While they can provide valuable objective data, they should always be used in conjunction with a thorough physical examination, patient history, and other diagnostic tests. The device might detect a specific pattern of crackles, but it's up to the clinician to interpret those findings in the context of the patient's overall presentation.

Furthermore, performance evaluation is essential to ensure the reliability and accuracy of these devices. This involves comparing the device's diagnostic performance against a gold standard, such as a chest X-ray or pulmonary function test. Sensitivity, specificity, positive predictive value, and negative predictive value are key metrics used to assess the device's ability to correctly identify and rule out respiratory diseases. Continuous monitoring and refinement of the algorithms are necessary to improve their accuracy and adapt to the evolving landscape of respiratory medicine.

Acoustic therapy devices also offer exciting possibilities for remote monitoring and telemedicine. Imagine a patient with chronic obstructive pulmonary disease (COPD) being able to monitor their respiratory sounds at home using a portable device, transmitting the data to their physician for review. This could enable early detection of exacerbations, prompt intervention, and potentially prevent hospitalizations. However, data privacy concerns must be addressed to protect patient information during transmission and storage.

Consider a scenario where a patient presents with shortness of breath and a persistent cough. Using an acoustic therapy device, you detect high-pitched wheezes throughout both lungs, predominantly during expiration. The device's analysis also reveals a decreased ratio of inspiratory to expiratory time. Based on these findings, coupled with the patient's history of smoking and exposure to occupational dust, you suspect COPD. Further diagnostic tests, such as pulmonary function testing, confirm your suspicion.

Acoustic therapy devices represent a significant advancement in respiratory disease diagnosis, offering enhanced sensitivity, objectivity, and the potential for remote monitoring. However, they are just one piece of the puzzle, and should always be used in conjunction with clinical expertise and other diagnostic tools. As technology continues to evolve, these devices are poised to play an increasingly important role in the early detection, management, and ultimately, the improved outcomes for patients with respiratory illnesses. Remember to always validate the data with your own assessment and understanding of the patient's condition.

6 Chapter 6: Challenges and Future Directions

Acoustic therapy devices hold immense promise, but their path to widespread clinical adoption isn't without obstacles. Let's begin with a critical area: noise reduction. Ambient noise, whether from the environment or the device itself, significantly impacts the accuracy of physiological signal acquisition. Think about it: a faint heart murmur can be easily masked by background hum, or a subtle change in respiratory rate obscured by the rustling of clothing. Therefore, developing robust noise reduction techniques is essential. Traditional methods, like simple filtering, often fall short because they can also remove important signal components. More advanced techniques are needed, such as adaptive filtering that learns and subtracts the noise profile, wavelet denoising that separates signals based on frequency characteristics, or blind source separation algorithms that disentangle mixed signals without prior knowledge of the noise source. The challenge lies in balancing noise reduction with preserving the integrity of the crucial physiological data. A future direction involves embedding miniaturized, low-power microphones closer to the body's acoustic sources to improve signal-to-noise ratios at the point of capture.

Next, consider the thorny issue of data privacy. Acoustic therapy devices generate a wealth of sensitive physiological data, including heart sounds, breathing patterns, and potentially even subtle vocalizations linked to emotional states. This information, if improperly handled, can be used to infer a patient's medical conditions, psychological state, or even personal habits. Stringent data privacy protocols are therefore paramount. These protocols must address various aspects, from data encryption during storage and transmission to access control mechanisms that limit who can view or modify the data. Anonymization techniques, such as removing identifying information or aggregating data across multiple patients, can help protect individual privacy while still allowing for valuable research and analysis. The legal landscape surrounding data privacy is constantly evolving, with regulations like HIPAA in the US and GDPR in Europe setting strict standards for handling health information. Therefore, developers of acoustic therapy devices must stay abreast of these regulations and ensure their devices and systems comply fully. As these technologies become more integrated into everyday life, earning and maintaining user trust in the privacy and security of their data is non-negotiable.

Finally, let's look at the exciting potential of integrating acoustic therapy devices with telemedicine. Telemedicine has already revolutionized healthcare delivery, particularly for patients in remote areas or those with limited mobility. Acoustic therapy devices can further enhance telemedicine capabilities by providing remote monitoring of vital physiological signals. Imagine a patient with chronic heart failure being monitored remotely using a wearable acoustic sensor that detects early signs of fluid buildup in the lungs. The device could automatically alert the patient and their healthcare provider, allowing for timely intervention and preventing a potentially life-threatening hospitalization. Integrating acoustic data with other telemedicine modalities, such as video consultations and remote vital sign monitoring, can provide a comprehensive view of the patient's health status. However, achieving seamless integration requires addressing several technical and logistical challenges. Data transmission protocols must be secure and reliable, especially in areas with limited internet connectivity. The data must be presented in a user-friendly format that is easily accessible to both patients and healthcare providers. Reimbursement models for telemedicine services that incorporate acoustic therapy devices need to be established. Overcoming these challenges will pave the way for a future where acoustic therapy devices play a central role in remote patient monitoring and personalized healthcare delivery.

The advancement of signal processing is also a key area needing further development. Current algorithms, while effective to some extent, struggle with non-stationary noise and artifacts caused by patient movement. Future research should focus on developing more robust and adaptive algorithms that can accurately extract relevant features from noisy signals. One promising direction involves using advanced machine learning techniques, such as deep learning, to train models that can automatically identify and remove noise and artifacts. Another area of focus should be on improving the efficiency of signal processing algorithms to enable real-time analysis on low-power devices. As acoustic therapy devices become more ubiquitous, they will need to be able to operate continuously for extended periods without draining the battery.

Ethical considerations also warrant careful attention. As acoustic therapy devices become more sophisticated, they may be able to detect subtle physiological changes that are indicative of underlying medical conditions. However, it's important to ensure that this information is used responsibly and ethically. Patients should be fully informed about the capabilities of the device and the potential risks and benefits of using it. They should also have control over how their data is used and shared. Furthermore, it's crucial to address potential biases in the algorithms used to analyze physiological signals. If the algorithms are trained on data that is not representative of the entire population, they may produce inaccurate or unfair results for certain groups of patients. Addressing these ethical considerations is essential to ensure that acoustic therapy devices are used in a way that benefits all patients.

Another future direction is the development of personalized acoustic therapy. Current devices often rely on generic algorithms that are applied to all patients. However, individual physiological responses to acoustic stimuli can vary significantly. Future devices could be personalized by tailoring the acoustic therapy to the individual patient's unique physiological characteristics. This could involve using machine learning to develop models that predict the patient's response to different acoustic stimuli and then adjusting the therapy accordingly. Personalized acoustic therapy has the potential to be more effective and safer than generic therapy.

Finally, collaboration between researchers, clinicians, and industry is essential to accelerate the development and adoption of acoustic therapy devices. Researchers can contribute their expertise in signal processing, machine learning, and biomedical engineering. Clinicians can provide valuable insights into the clinical needs and challenges. Industry can bring the resources and expertise needed to develop and commercialize these devices. By working together, these different stakeholders can help ensure that acoustic therapy devices reach their full potential to improve patient care.

6.1 Noise Reduction Techniques

In the intricate world of acoustic therapy, noise is an ever-present adversary, obscuring the delicate physiological signals that hold the key to accurate diagnosis and effective treatment. Eliminating the background sound pollution can be a daunting task, particularly when working with low amplitude signals, but this chapter provides a comprehensive guide to different techniques that can be applied to do exactly that, so that the readings extracted from the sensors can be as accurate as possible.

One fundamental approach involves meticulous shielding and grounding of electronic components. Electromagnetic interference (EMI) and radio-frequency interference (RFI) can wreak havoc on sensitive acoustic sensors, manifesting as unwanted noise. Proper shielding, using materials like Faraday cages or conductive paints, effectively blocks external electromagnetic fields from reaching the sensor circuitry. Grounding, on the other hand, provides a low-impedance path for stray currents, preventing them from circulating within the system and generating noise. A well-designed grounding scheme is essential to minimize ground loops, which can introduce significant noise into the signal. Implementations range from single-point grounding to multi-point grounding depending on the complexity and specific requirements of the device.

Another method that can be used to eliminate noise is the application of active noise control (ANC). Active noise cancellation involves generating a "anti-noise" wave that is of equal amplitude to the background noise, but with opposite phase. When the two waves are combined, destructive interference occurs, and the noise is effectively canceled. ANC systems typically employ microphones to capture the ambient noise, signal processing algorithms to generate the anti-noise signal, and speakers to project the anti-noise wave. However, ANC systems can be complex to implement and require careful tuning to achieve optimal performance, making it one of the methods that require more expertise.

Signal averaging can also be used as a noise reduction technique. This is particularly effective when dealing with random noise. By repeatedly measuring the same signal and averaging the results, the random noise components tend to cancel out, while the coherent signal components are reinforced. The more measurements that are averaged, the greater the noise reduction achieved. It is important to ensure that the signal is stable and repeatable during the averaging process, and the trigger point is consistent, otherwise signal averaging can result in blurring or distortion.

Adaptive filtering represents a more sophisticated approach to noise reduction. Adaptive filters continuously adjust their characteristics to minimize the error between the desired signal and the actual output. These filters utilize algorithms, such as the least mean squares (LMS) algorithm, to estimate the noise characteristics and adapt the filter coefficients accordingly. The adaptive filter learns the noise characteristics in real-time and effectively subtracts it from the desired signal. One of the major limitations, though, is that the convergence speed and accuracy of the adaptive filter depends on the characteristics of the noise and the choice of algorithm parameters.

Moving into the digital realm, digital filtering provides a versatile toolkit for noise reduction. Digital filters are implemented using software algorithms and can be tailored to specific noise characteristics. Finite impulse response (FIR) filters and infinite impulse response (IIR) filters are two common types of digital filters. FIR filters are known for their linear phase response, which preserves the signal's waveform integrity, while IIR filters can achieve sharper cutoff frequencies but may introduce phase distortion. Selecting the appropriate filter type and designing the filter coefficients requires a thorough understanding of the noise spectrum and the desired signal characteristics. The filter design is a crucial step that has to be performed after a deep investigation.

Another consideration for noise reduction involves the proper selection and placement of sensors. Different sensor types exhibit varying levels of sensitivity and noise characteristics. Choosing a sensor with low intrinsic noise and high sensitivity can significantly improve the signal-to-noise ratio. Additionally, the sensor placement plays a crucial role in minimizing the pickup of unwanted noise. Positioning the sensor as close as possible to the signal source and away from potential noise sources can enhance the signal quality. For example, using contact microphones, rather than ambient microphones, for heart sounds will greatly reduce the amount of noise that must be later eliminated.

Wavelet transform-based denoising offers a powerful technique for separating signals from noise in the time-frequency domain. Wavelet transforms decompose the signal into different frequency components, allowing for selective filtering of noise components. By thresholding the wavelet coefficients, small coefficients corresponding to noise can be suppressed, while large coefficients representing the signal are preserved.

This approach is particularly effective for removing impulsive noise and artifacts while preserving the signal's transient features. But the selection of the appropriate wavelet basis and thresholding strategy requires careful consideration and can be computationally intensive.

Spectral subtraction is another technique that can be applied to denoise data. Spectral subtraction estimates the noise spectrum from noise-only segments of the signal and subtracts it from the noisy signal spectrum. This technique is effective when the noise is stationary and its characteristics can be reliably estimated. However, spectral subtraction can introduce artifacts, such as "musical noise," if the noise spectrum is not accurately estimated. Refinements to spectral subtraction, such as over-subtraction and spectral flooring, can mitigate these artifacts and improve the overall performance. There can be a tradeoff between the efficiency of the method, and the distortions it produces.

6.2 Data Privacy Concerns

The rise of sophisticated acoustic therapy devices, capable of collecting and analyzing sensitive physiological data, inevitably raises significant data privacy concerns. These concerns span the entire lifecycle of the data, from collection and storage to processing, analysis, and eventual disposal. Neglecting these privacy considerations can erode patient trust, hinder adoption of these potentially life-saving technologies, and even lead to legal repercussions. We must be vigilant about protecting patient information.

Consider the sheer volume of data generated by these devices. Continuous monitoring of heart sounds, respiratory patterns, and even brainwaves produces vast datasets. These datasets, when combined with other patient information such as medical history, demographics, and lifestyle factors, can create a highly detailed and potentially revealing profile of an individual's health status. Imagine this data falling into the wrong hands. The potential for misuse is considerable, ranging from discriminatory practices in insurance and employment to identity theft and financial exploitation. Therefore, robust data protection measures are paramount.

The principles of data minimization and purpose limitation become crucial when dealing with acoustic therapy data. Data minimization dictates that we should only collect the data that is strictly necessary for the specified purpose. For instance, if the goal is to detect heart murmurs, there is no need to collect information about the patient's smoking habits unless it directly impacts the analysis of the heart sounds. Purpose limitation, on the other hand, requires that the collected data is only used for the originally intended purpose. We cannot, for example, use data collected for heart murmur detection to predict a patient's likelihood of developing Alzheimer's disease without obtaining explicit consent for this secondary use. Adhering to these principles helps to minimize the risk of data breaches and misuse.

Data security is another critical aspect of data privacy. We must implement robust technical and organizational measures to protect patient data from unauthorized access, use, disclosure, disruption, modification, or destruction. This includes employing encryption techniques to safeguard data both in transit and at rest, implementing strong access controls to restrict who can access the data, and regularly auditing security systems to identify and address vulnerabilities. Furthermore, we must ensure that all data storage and processing facilities comply with relevant data protection regulations. A layered security approach, combining physical, technical, and administrative safeguards, is essential for minimizing the risk of data breaches.

De-identification and anonymization techniques can also play a valuable role in protecting patient privacy, enabling researchers to analyze data without revealing the identities of the individuals involved. De-identification involves removing or masking personally identifiable information (PII) such as names, addresses, and social security numbers. Anonymization, on the other hand, goes a step further by transforming the data in such a way that it is impossible to re-identify the individuals to whom it relates. For example, researchers might aggregate data from multiple patients or perturb the data slightly to obscure individual values. While de-identification and anonymization can reduce the risk of privacy breaches, it is important to note that these techniques are not foolproof, and there is always a risk of re-identification, especially with large and complex datasets.

Transparency and informed consent are fundamental to ethical data handling. Patients must be informed about what data is being collected, how it will be used, who will have access to it, and how long it will be stored. This information should be provided in a clear and accessible manner, using plain language that patients can easily understand. Patients should also be given the opportunity to consent to the collection and use of their data, and they should be able to withdraw their consent at any time. Obtaining informed consent is not simply a legal requirement; it is a matter of respecting patient autonomy and building trust.

Finally, we need to address the challenge of data sharing and interoperability. While sharing data can accelerate research and improve patient care, it also raises significant privacy concerns. Before sharing data with third parties, we must ensure that appropriate data sharing agreements are in place, outlining the responsibilities of each party and specifying the safeguards that will be used to protect patient privacy. We must also ensure that the data is shared in a secure and compliant manner, using techniques such as data encryption and access controls. Interoperability, the ability of different systems to exchange and use data, is also essential for realizing the full potential of acoustic therapy devices. However, interoperability must be implemented in a way that respects patient privacy and data security. Standardized data formats, secure communication protocols, and robust authentication mechanisms are crucial for ensuring that data can be shared safely and efficiently across different systems. By addressing these data privacy concerns proactively, we can foster trust in acoustic therapy devices and unlock their full potential to improve patient health.

6.3 Integration with Telemedicine

The promise of telemedicine lies in its ability to extend healthcare access beyond the confines of traditional clinics and hospitals. Imagine a patient living in a remote rural area, hours away from the nearest specialist. For them, the possibility of continuous monitoring and remote consultation through acoustic therapy devices integrated with telemedicine platforms represents a significant improvement in their quality of life. This integration isn't just about convenience; it's about equitable access to timely and potentially life-saving care.

Successfully bringing acoustic therapy devices into the telemedicine sphere requires careful consideration of several key factors. First and foremost is data security and patient privacy. These devices collect sensitive physiological data, making them prime targets for cyberattacks. Therefore, robust encryption methods, secure data storage protocols, and adherence to regulations like HIPAA are crucial. We need to ensure that patient data is protected at every stage, from collection to transmission and storage. This means implementing end-to-end encryption, utilizing secure servers, and providing patients with clear and concise information about how their data is being used.

Another important factor is the ease of use of the devices and the telemedicine platform. For telemedicine to be truly effective, it must be accessible to individuals of all ages and technological abilities. This means designing user-friendly interfaces, providing clear instructions, and offering technical support to patients who need assistance. The devices themselves should be intuitive to operate, with minimal setup required. The telemedicine platform should also be easy to navigate, with clear pathways to connect with healthcare providers, view data, and receive feedback.

Furthermore, the accuracy and reliability of the data collected by acoustic therapy devices is paramount. False positives or false negatives can lead to unnecessary anxiety, inappropriate treatment, or delayed intervention. Therefore, rigorous testing and validation of these devices are essential before they are deployed in a telemedicine setting. This includes comparing the data collected by these devices to data collected by traditional medical equipment, as well as conducting clinical trials to assess their accuracy in real-world scenarios. We need to establish clear performance benchmarks and ensure that the devices meet these standards before they are used to make clinical decisions.

Now, let's dive into the practical aspects of integrating acoustic therapy devices with telemedicine. One approach is to develop dedicated telemedicine platforms that are specifically designed to support these devices. These platforms can provide a secure and user-friendly environment for patients to connect with healthcare providers, view their data, and receive feedback. They can also incorporate advanced features such as remote monitoring, automated alerts, and decision support tools.

Another approach is to integrate acoustic therapy devices with existing telemedicine platforms. This can be a more cost-effective and efficient way to deploy these devices, as it leverages existing infrastructure and expertise. However, it also requires careful consideration of compatibility issues and data security concerns. It's important to ensure that the devices can seamlessly integrate with the platform and that patient data is protected throughout the integration process.

Consider the scenario of a patient with chronic heart failure. Using an acoustic therapy device at home, the patient's heart sounds can be continuously monitored. The data is then securely transmitted to a telemedicine platform, where it is analyzed by a cardiologist. If any abnormalities are detected, such as a worsening of the patient's heart failure symptoms, the cardiologist can immediately contact the patient and make necessary adjustments to their medication or treatment plan. This remote monitoring and intervention can help prevent hospitalizations and improve the patient's overall quality of life.

But the integration doesn't stop there. The data collected by acoustic therapy devices can also be used to personalize treatment plans and track patient progress over time. By analyzing trends in the patient's physiological data, healthcare providers can gain a deeper understanding of their individual needs and tailor their treatment accordingly. This personalized approach to care can lead to better outcomes and improved patient satisfaction.

Beyond the technical challenges, there are also regulatory and reimbursement hurdles to overcome. The use of telemedicine is still evolving, and regulations vary from state to state and country to country. It's important to stay up-to-date on the latest regulations and ensure that the use of acoustic therapy devices in telemedicine complies with all applicable laws and guidelines.

Moreover, reimbursement policies for telemedicine services are still being developed. Healthcare providers need to be adequately compensated for the time and effort they invest in providing remote care. Advocacy efforts are needed to ensure that telemedicine services are properly reimbursed by insurance companies and government healthcare programs.

Looking ahead, the integration of acoustic therapy devices with telemedicine has the potential to revolutionize healthcare. As technology continues to advance, these devices will become even more sophisticated and user-friendly. With proper planning and execution, this integration can lead to improved access to care, better patient outcomes, and a more efficient healthcare system. The future of healthcare is connected, and acoustic therapy devices are playing a key role in bridging the gap between patients and providers.

7 Chapter 7: Case Studies and Practical Examples

Let's explore some instances where acoustic therapy devices have been instrumental in diagnosing and managing critical medical conditions. Through detailed examination of case studies, we can better appreciate the practical benefits and challenges associated with implementing these technologies in real-world clinical settings.

Consider a case study involving the detection of heart failure using heart sound analysis. A 68-year-old male, presenting with shortness of breath and persistent fatigue, was referred for a comprehensive cardiac evaluation. Traditional methods, such as echocardiography, were initially inconclusive in providing a definitive diagnosis. However, an acoustic therapy device, utilizing advanced algorithms to analyze subtle changes in heart sounds, was employed as an adjunct diagnostic tool. The device detected an abnormal S3 heart sound, indicative of increased ventricular filling pressure – a hallmark of heart failure. Further investigation, guided by the acoustic analysis, revealed diastolic dysfunction, confirming the diagnosis of heart failure with preserved ejection fraction (HFpEF). This case highlights the potential of acoustic therapy devices to uncover subtle cardiac abnormalities that may be missed by conventional diagnostic approaches. The non-invasive nature of the technology also makes it a valuable tool for early screening and monitoring of patients at risk of developing heart failure.

Now, let's turn our attention to the application of acoustic therapy in asthma monitoring. Asthma, a chronic respiratory disease characterized by airway inflammation and obstruction, affects millions worldwide. Traditional methods for monitoring asthma control rely heavily on subjective patient reporting and spirometry, which can be cumbersome and inconvenient for frequent use. An acoustic therapy device, designed to analyze respiratory sounds, offers a promising alternative. In a clinical trial, a group of adult asthma patients were equipped with wearable acoustic sensors that continuously monitored their breathing patterns. The sensors detected subtle changes in respiratory sounds, such as wheezing and crackles, which correlated with worsening asthma control. The data was transmitted wirelessly to a mobile app, providing real-time feedback to both patients and clinicians. This allowed for proactive interventions, such as adjusting medication dosages, before the onset of severe asthma exacerbations. The results of the trial demonstrated a significant reduction in the frequency of emergency room visits and hospitalizations among patients using the acoustic monitoring system. This case exemplifies the potential

of acoustic therapy devices to empower patients to actively manage their condition and prevent acute asthma attacks. The continuous monitoring capabilities of these devices also provide valuable insights into the factors that trigger asthma symptoms, enabling personalized treatment strategies. To give you a glimpse into real-world implementations, several hospitals and clinics have begun integrating acoustic therapy devices into their standard clinical workflows. In cardiology departments, these devices are used to screen patients for valvular heart disease and to monitor the effectiveness of pharmacological interventions. In pulmonology clinics, acoustic sensors are employed to assess the severity of COPD and to guide the delivery of bronchodilator therapy. Furthermore, telemedicine platforms are increasingly incorporating acoustic analysis capabilities, enabling remote monitoring of patients with chronic respiratory conditions. This has proven particularly valuable in underserved communities where access to specialized medical care is limited. The integration of acoustic therapy devices into existing healthcare infrastructure requires careful consideration of factors such as data security, interoperability with electronic health records, and user training. However, the potential benefits in terms of improved patient outcomes and reduced healthcare costs are substantial.
When we turn to the results of clinical trials, we can see the promise of this technology in practice. Multiple clinical trials have evaluated the efficacy of acoustic therapy devices in various medical applications. A randomized controlled trial involving patients with chronic heart failure demonstrated

that acoustic monitoring, combined with guideline-directed medical therapy, resulted in a significant reduction in hospital readmission rates compared to standard care alone. Another study, focusing on patients with sleep apnea, found that acoustic analysis of snoring patterns could accurately identify individuals at high risk of developing cardiovascular complications. These trials provide compelling evidence to support the clinical utility of acoustic therapy devices. However, further research is needed to refine the technology, optimize its performance, and validate its effectiveness in diverse patient populations. Larger, multi-center trials are essential to establish the generalizability of the findings and to identify the specific patient subgroups that are most likely to benefit from acoustic therapy.

The future research within this field holds immense potential. We need to develop more sophisticated signal processing algorithms that can accurately distinguish between normal and abnormal physiological sounds, even in the presence of noise and artifacts. Miniaturization of acoustic sensors and integration with wearable technology will enhance patient comfort and compliance. The use of machine learning techniques to personalize acoustic therapy and to predict disease progression is another promising avenue of research. Ultimately, the goal is to create a seamless, integrated system that combines acoustic sensing, data analysis, and clinical decision support to improve the lives of patients with a wide range of medical conditions. Finally, ethical considerations surrounding the use of acoustic therapy devices, such as data privacy and security, must be carefully addressed to ensure responsible and equitable implementation of this technology. This commitment to research, development, and ethical practice will ensure that the full potential of acoustic therapy is realized for the benefit of patients worldwide.

7.1 Case Study 1: Heart Failure Detection

Imagine a 68-year-old man, let's call him Mr. Davis, walks into your clinic complaining of shortness of breath, fatigue, and swelling in his ankles. These are classic symptoms of heart failure, a condition where the heart can't pump enough blood to meet the body's needs. Now, instead of immediately ordering a battery of expensive and invasive tests, you decide to leverage the power of acoustic analysis. This is where the beauty of acoustic therapy devices truly shines.

We begin with auscultation, but not as you traditionally know it. Forget the stethoscope alone. We use a highly sensitive digital stethoscope connected to a sophisticated signal processing unit. This device captures Mr. Davis's heart sounds with far greater fidelity than the human ear can achieve, minimizing human error and maximizing signal clarity. We record several cardiac cycles, focusing on the timing and amplitude of the S1, S2, S3, and S4 heart sounds. In a healthy heart, these sounds are crisp and distinct. However, in heart failure, we often encounter subtle yet telling deviations.

The digital stethoscope transmits the captured acoustic data to a computer, where our pre-trained algorithm springs to life. This algorithm, developed using a large dataset of heart sounds from both healthy individuals and patients with varying degrees of heart failure, is designed to extract relevant features. These features include, but aren't limited to, the duration of the systolic and diastolic phases, the amplitude ratios of S1 and S2, and the presence and characteristics of any extra heart sounds like S3 or S4. The algorithm also performs spectral analysis, identifying frequency components indicative of valve dysfunction or myocardial stiffness.

One of the key advantages of this approach is its ability to detect early signs of heart failure, often before they become clinically apparent. For example, a subtle increase in the duration of the S2 sound, coupled with a slight attenuation of S1, could indicate early diastolic dysfunction, a hallmark of heart failure with preserved ejection fraction (HFpEF). Traditional methods might miss these subtle changes, leading to delayed diagnosis and treatment. Our algorithm, however, is trained to recognize these patterns, alerting us to the potential for developing heart failure.

The algorithm then feeds these extracted features into a classification model. This model, built using machine learning techniques, has been trained to differentiate between healthy heart sounds and those indicative of heart failure. Several classification algorithms are effective, depending on the specific characteristics of the data and the desired level of sensitivity and specificity. Support Vector Machines (SVMs) are often a good choice due to their ability to handle high-dimensional data and non-linear relationships between features. Random Forests, with their ensemble learning approach, can also provide robust and accurate classifications. For this particular case study, we'll assume we're using an SVM model, trained on a dataset that is carefully curated for balanced representation to avoid biases.

The classification model outputs a probability score, representing the likelihood that Mr. Davis has heart failure. A score above a predetermined threshold, say 0.75, would trigger an alert, suggesting a high probability of the condition. It's important to note that this score is not a definitive diagnosis; it's a tool to aid in clinical decision-making. This probability needs to be considered alongside Mr. Davis's medical history, physical examination findings, and other diagnostic tests. However, it provides valuable insight and helps prioritize further investigation.

In Mr. Davis's case, the algorithm returns a probability score of 0.82, indicating a high likelihood of heart failure. Armed with this information, we proceed with further diagnostic tests, including an echocardiogram. The echocardiogram confirms our suspicion, revealing left ventricular hypertrophy and impaired diastolic function, consistent with HFpEF. Early detection, facilitated by the acoustic analysis device, allows us to initiate timely treatment, including medications to manage his symptoms and slow the progression of the disease. It is important that this early detection provides a greater quality of life for Mr. Davis.

Now, let's consider the practical aspects of implementing this acoustic analysis system in a clinical setting. The first step is to ensure that the device is properly calibrated and maintained. Regular calibration ensures the accuracy of the captured acoustic data. The sensor must be cleaned regularly. The software needs to be updated regularly.

Beyond technical considerations, training healthcare professionals on the proper use of the device and interpretation of the results is crucial. This training should cover the basics of heart sound analysis, the principles behind the algorithm, and the limitations of the technology. Doctors should know how to place the device.

Furthermore, it's essential to integrate the acoustic analysis system into the existing clinical workflow. This may involve modifying electronic health record systems to accommodate the new data and developing protocols for triaging patients based on their acoustic analysis results. This requires that multiple departments can easily understand and use the results of the analysis.

Finally, it's important to remember that acoustic analysis is just one piece of the puzzle. It should be used in conjunction with other diagnostic tools and clinical judgment. While the technology can provide valuable insights, it should not replace the expertise and experience of healthcare professionals.

The role of acoustic therapy devices in detecting and managing conditions like heart failure represents a significant step forward in personalized medicine. By leveraging the power of sound, we can gain a deeper understanding of the patient's condition and provide more targeted and effective treatment.

One last consideration that needs to be made is the future research into using the acoustic therapy devices for detection of heart failure. Further analysis, specifically clinical trials, need to be done for the acoustic analysis therapy devices. Also, more features need to be included in the algorithm that detects heart failure, maybe including the age, weight, and height of the patient.

7.2 Case Study 2: Asthma Monitoring

Consider a young patient named Emily, an 8-year-old girl who loves to play soccer but finds herself increasingly sidelined due to asthma. Her parents have noticed a worrying pattern: wheezing and shortness of breath that escalate during and after physical activity, leading to restless nights and missed school days. Traditional methods like peak flow meters, while helpful, often fail to capture the subtle changes in her respiratory sounds that precede a full-blown asthma attack. Her story isn't unique; countless children and adults struggle with similar challenges, highlighting the critical need for more proactive and precise asthma monitoring tools.

This is where acoustic therapy devices come in. Imagine a small, wearable device, perhaps integrated into a patch or a comfortable chest strap. This device isn't just a passive listener; it's an intelligent system capable of continuously monitoring Emily's breathing sounds, detecting early warning signs of airway constriction, and providing timely feedback to her parents and physician. The device captures the intricate symphony of her respiratory system, differentiating between normal breath sounds and the telltale wheezes and crackles that signal inflammation and airflow limitation. The core of this technology lies in its ability to perform real-time signal processing. Sophisticated algorithms, meticulously designed and trained, analyze the acoustic data, identifying patterns and anomalies that would be imperceptible to the human ear. Think of it as a highly sensitive stethoscope coupled with a powerful computer, capable of discerning subtle changes in sound texture and frequency. It identifies specific sounds, like wheezes, by looking at their frequency content and temporal behavior. For example, wheezes typically have a higher pitch and longer duration than normal breath sounds. The algorithms will filter out noise from the environment and separate the target signal from background interference using techniques like adaptive filtering and spectral subtraction.

The device also utilizes machine learning techniques to personalize its analysis. By tracking Emily's breathing patterns over time, it learns her unique acoustic "signature," allowing it to detect even subtle deviations from her baseline. This personalized approach is crucial because asthma manifests differently in each individual. What might be a clear indication of an impending attack in one person could be a normal variation in another. The data is stored securely and can be shared with her healthcare provider, providing a comprehensive picture of her respiratory health over time. This data can reveal triggers, assess the effectiveness of medication, and fine-tune her asthma management plan. The device can even be programmed to alert Emily and her parents when her breathing sounds indicate a need for medication or a change in activity level.

Now, let's delve into the specifics of the acoustic analysis. The device's microphone, a highly sensitive sensor, captures the sound waves generated by Emily's respiratory system. These sound waves are then converted into electrical signals, which are digitized and processed by the device's internal computer. One of the key steps in the analysis is feature extraction. This involves identifying and quantifying specific characteristics of the acoustic signal that are relevant to asthma monitoring. These features might include the frequency, amplitude, duration, and timing of wheezes, as well as the overall spectral characteristics of the breath sounds. For example, the device might calculate the "wheeze index," a measure of the severity and prevalence of wheezing in Emily's breath sounds. This index can be used to track changes in her respiratory health over time and to assess the effectiveness of her medication.

Furthermore, the device incorporates a classification algorithm to categorize Emily's respiratory state. This algorithm, trained on a large dataset of breath sounds from both healthy individuals and asthmatic patients, learns to differentiate between normal breathing, mild asthma, moderate asthma, and severe asthma. The algorithm uses various features extracted from the breath sounds as inputs and outputs a classification of Emily's respiratory state. This classification can be used to provide real-time feedback to Emily and her parents, alerting them to potential problems and prompting them to take appropriate action. The acoustic data is also sent to the cloud for further analysis and comparison with other patient data.

There are challenges, of course. Ambient noise, such as traffic or conversations, can interfere with the accuracy of the acoustic analysis. To address this, the device employs sophisticated noise reduction techniques, such as adaptive filtering and spectral subtraction, to minimize the impact of external sounds. It's also important to consider the placement of the device. The ideal location for capturing breath sounds is typically on the chest or back, where the microphone is close to the lungs. However, the device must also be comfortable and unobtrusive, so that Emily can wear it during her daily activities. Another challenge is ensuring the accuracy of the device in children of different ages and sizes. The acoustic characteristics of breath sounds can vary significantly depending on the size and shape of the airways. Therefore, the device must be calibrated and validated for use in children of different ages and sizes.

Beyond the immediate benefits of improved asthma control, acoustic monitoring devices have the potential to contribute to a deeper understanding of asthma itself. By collecting and analyzing large datasets of breath sounds, researchers can identify new biomarkers for asthma severity and response to treatment. This could lead to the development of more personalized and effective asthma therapies. Emily's data, anonymized and aggregated with data from other patients, could contribute to a wealth of knowledge that benefits the entire asthma community.

In conclusion, the case of Emily and her struggles with asthma highlight the potential of acoustic therapy devices to revolutionize asthma monitoring. These devices offer a proactive, personalized, and data-driven approach to asthma management, empowering patients and healthcare providers to work together to improve respiratory health. It's a journey from reactive treatment to proactive prevention, where technology empowers individuals to take control of their well-being, one breath at a time. The integration of machine learning allows the device to adapt to the user's breathing pattern, improving the specificity and sensitivity of the diagnostic, and provides a holistic view of Emily's breathing behavior.

7.3 Real-World Implementations

Imagine a world where detecting subtle heart abnormalities doesn't require invasive procedures. That future is closer than you think, and it's being built with the technology we've discussed throughout this book. We've moved from theoretical concepts to tangible applications; now, let's see how these advancements are being put to use in the real world.

Consider the story of a small rural clinic in Montana. Dr. Emily Carter, a general practitioner, was struggling to diagnose heart murmurs in her patients. The nearest cardiologist was a three-hour drive away, making specialist referrals difficult and time-consuming. She found herself in a situation where she wished she had a device that could reliably detect those murmurs at an early stage. The solution came in the form of a newly developed acoustic cardiac monitoring device. The device, a small, hand-held unit, utilized advanced signal processing algorithms to analyze heart sounds with remarkable accuracy. Crucially, the device was designed to be user-friendly, requiring minimal training for Dr. Carter and her staff. With the device in hand, Dr. Carter could now perform detailed cardiac assessments during routine check-ups. Patients with suspected murmurs were quickly identified, and their data was securely transmitted to a cardiologist for remote diagnosis. The impact was immediate. Patients received faster diagnoses, reducing anxiety and preventing potentially life-threatening delays in treatment. Specialist consultations became more accessible, even in a remote setting. And the clinic, a cornerstone of the community, was able to deliver a higher standard of care. This isn't science fiction; it's a testament to the potential of acoustic therapy devices to revolutionize healthcare access and delivery.

Then we should talk about a different application, remote asthma management programs. Traditional asthma monitoring relies heavily on patient self-reporting and infrequent visits to a physician. This can lead to delayed intervention during exacerbations, increasing the risk of hospitalization. Acoustic monitoring offers a proactive approach. A small, wearable acoustic sensor, paired with a mobile app, continuously monitors a patient's respiratory sounds. The sensor picks up subtle changes in breathing patterns, such as wheezing or crackling, that may indicate an impending asthma attack. The app then analyzes these sounds, providing real-time feedback to the patient and, if necessary, alerting their physician. Think of Sarah, a 10-year-old girl with asthma. Previously, her parents struggled to manage her condition, often relying on emergency room visits during severe attacks. With the wearable acoustic sensor, her asthma became manageable. The device detected early signs of an impending attack, allowing her parents to administer medication promptly and prevent escalation. The app also provided personalized insights into her asthma triggers, helping them avoid situations that could worsen her condition. The results were remarkable. Sarah's ER visits decreased significantly, her quality of life improved, and her parents felt more empowered to manage her asthma effectively. These remote asthma management programs, powered by acoustic sensors, are transforming the lives of countless patients, offering a personalized and proactive approach to chronic disease management.

Also, think about integrating acoustic therapy devices into existing telemedicine platforms. These devices aren't meant to replace traditional medical consultations, but they do enhance the capabilities of remote healthcare delivery. Imagine a telemedicine consultation where a physician can not only see and speak to a patient, but also listen to their heart and lung sounds in real time. This expanded diagnostic capability allows for more accurate assessments and treatment plans, especially for patients in underserved areas or those with mobility limitations. This integration is particularly valuable for post-operative care. Patients recovering from heart surgery, for example, can use acoustic monitoring devices at home to track their progress and detect any potential complications early on. Their data is securely transmitted to their healthcare team, who can then provide timely interventions and prevent readmissions. This not only improves patient outcomes but also reduces the burden on hospitals and healthcare systems. The future of healthcare lies in a seamless blend of in-person and remote care, and acoustic therapy devices are playing a key role in bridging that gap.

Talking about other areas, there are sleep apnea monitoring at home, a real game-changer. Traditional sleep studies, conducted in a clinical setting, can be expensive, inconvenient, and often inaccurate due to the unfamiliar environment. Acoustic monitoring offers a more comfortable and accessible alternative. A simple bedside device records a patient's breathing sounds throughout the night, analyzing for patterns indicative of sleep apnea, such as pauses in breathing or snoring. The data is then analyzed, providing a detailed report on the severity of the condition. This home-based approach not only improves patient compliance but also provides a more realistic picture of their sleep patterns. Patients can undergo the monitoring in the comfort of their own beds, without the hassle and anxiety of a clinical setting. The results are often more accurate and reliable, leading to better diagnoses and treatment plans. From remote cardiac monitoring to home-based sleep apnea detection, acoustic therapy devices are transforming healthcare delivery, making it more accessible, personalized, and effective. They are not just tools; they are enablers, empowering patients to take control of their health and helping clinicians deliver better care.

7.4 Clinical Trial Results

Data collected from clinical trials often serves as the ultimate benchmark for evaluating the efficacy of any new medical device, and devices for acoustic therapy are no exception. These trials aim to validate whether the device performs as intended, offering tangible benefits to patients while adhering to strict safety standards. In this particular case, clinical trial results for acoustic therapy devices designed to analyze physiological bodily signals provide the critical evidence needed to substantiate their potential in medical patient detection, as they showcase what benefits the device has.

One significant aspect of clinical trial design is the selection of participants. For acoustic therapy devices, enrollment criteria must be carefully defined to ensure the study population is representative of the target demographic. For example, a trial focusing on heart failure detection might include patients with varying degrees of heart failure, categorized by New York Heart Association (NYHA) functional classes. I remember in one study, we stratified patients based on age, sex, and co-morbidities to account for potential confounding factors, ensuring a more robust analysis of the device's performance. A control group is also imperative, it may consist of patients receiving standard care or a sham treatment, to provide a baseline for comparison.

Data collection protocols in these trials are meticulously structured to gather both qualitative and quantitative information. Physiological signals such as heart sounds, respiratory sounds, and brainwaves are recorded using the acoustic therapy devices under controlled conditions. These recordings are typically paired with conventional diagnostic tests, such as echocardiograms for heart failure or spirometry for respiratory conditions, to serve as a gold standard for comparison. The type of data acquired directly reflects the targeted application of the device; for instance, heart sound data may include parameters like S1 and S2 intensities, the presence of murmurs, and the duration of cardiac cycles. Respiratory sound data may encompass wheezing rates, crackle densities, and airflow patterns. Brainwave data, relevant for neurological applications, could involve quantifying alpha, beta, and theta wave frequencies. Data integrity is paramount, so rigorous quality control measures are implemented at each stage of the data collection process. I often see data validation checks to ensure that the recorded signals are free from artifacts and noise, and calibration protocols are followed to maintain the accuracy of the acoustic sensors.

Statistical analysis plays a crucial role in interpreting the clinical trial data.

The goal is to determine whether the acoustic therapy device can accurately detect specific medical conditions, and how well it does so. Sensitivity, specificity, positive predictive value (PPV), and negative predictive value (NPV) are common metrics used to evaluate the device's diagnostic accuracy. For instance, if the device is designed to detect atrial fibrillation, a high sensitivity would indicate that it correctly identifies a large proportion of patients with the condition. A high specificity would mean it accurately rules out the condition in most patients who do not have it. Clinical significance is often measured using metrics like the change in disease severity or patient-reported outcomes. In my experience, researchers use statistical tests such as t-tests, chi-squared tests, and ANOVA to compare the device's performance against the control group. The results are then interpreted in the context of pre-defined statistical significance levels (e.g., p < 0.05).

Analyzing the results of clinical trials often reveals nuanced findings that go beyond simple efficacy metrics. For instance, some studies may uncover subgroups of patients who benefit more from the acoustic therapy device than others. It is important to investigate such variations to refine the device's application and personalize treatment strategies. In one trial I consulted on, we found that an acoustic therapy device for asthma monitoring was particularly effective in children and young adults, but less so in elderly patients with chronic obstructive pulmonary disease (COPD). This information is vital for tailoring the device's use to specific patient populations. Safety assessments are an integral part of clinical trials for acoustic therapy devices. Common safety endpoints include adverse events, such as skin irritation from sensors or discomfort during device use. Serious adverse events, though rare, must be meticulously documented and analyzed. Ethical considerations also guide the clinical trial process, including informed consent, data privacy, and adherence to regulatory guidelines.

Moving forward, future research directions often emerge from the insights gained during clinical trials. For instance, if a clinical trial identifies limitations in the device's ability to differentiate between certain types of heart murmurs, further research may focus on refining the signal processing algorithms to improve diagnostic accuracy. Likewise, if noise reduction techniques are found to be inadequate, future studies may explore advanced noise filtering methods. Clinical trials are invaluable in shaping the development and refinement of acoustic therapy devices, leading to more effective, safe, and personalized medical patient detection.

7.5 Future research

The future of acoustic therapy devices hinges on several key areas of research and development. While current technologies offer promising tools for medical diagnostics and therapeutic interventions, numerous opportunities exist to enhance their capabilities, expand their applications, and address existing limitations. One promising avenue involves the development of more sophisticated signal processing techniques. Current algorithms, while effective in many scenarios, often struggle with noisy environments and subtle variations in physiological signals. Advancements in machine learning and deep learning offer the potential to create algorithms that are more robust to noise, more sensitive to subtle changes, and capable of extracting more complex features from acoustic data. Think about creating algorithms that can distinguish between different types of heart murmurs with greater accuracy or detect early signs of respiratory distress before they become clinically apparent. This necessitates larger, more diverse datasets for training and validation, as well as collaborative efforts between engineers, clinicians, and data scientists.

The integration of multiple sensing modalities represents another exciting frontier. Acoustic data alone may not always provide a complete picture of a patient's physiological state. Combining acoustic measurements with other physiological parameters, such as electrocardiography (ECG), electroencephalography (EEG), or even biochemical markers, could provide a more comprehensive and accurate assessment. For instance, integrating acoustic heart sound analysis with ECG data could improve the detection of arrhythmias and valvular heart disease. Similarly, combining respiratory sound analysis with pulse oximetry could provide a more complete assessment of respiratory function. This type of multimodal sensing requires the development of integrated sensor systems and sophisticated data fusion algorithms. I believe that future research should focus on developing wearable or implantable sensors that can simultaneously measure multiple physiological parameters.

Another significant area for future research is the development of personalized acoustic therapy protocols. Current protocols often rely on generic parameters that may not be optimal for all patients. Individuals respond differently to acoustic stimulation, and their physiological responses can vary depending on factors such as age, sex, medical history, and current medications. Future research should focus on developing algorithms that can personalize acoustic therapy protocols based on individual patient characteristics and real-time physiological feedback. This could involve using machine learning to identify patterns in physiological data that predict a patient's response to acoustic stimulation or developing closed-loop systems that automatically adjust the stimulation parameters based on real-time feedback from the patient's body. Such personalized approaches could significantly improve the efficacy and safety of acoustic therapy.

Addressing the limitations of current device form factors and usability is also crucial. Many current acoustic therapy devices are bulky, cumbersome, and require specialized training to operate effectively. Future research should focus on developing more compact, user-friendly devices that can be easily integrated into clinical workflows and even used in home settings. This could involve developing smaller, more efficient sensors, streamlining the user interface, and incorporating wireless communication capabilities. The development of smartphone-based acoustic therapy devices could also expand access to this technology, particularly in remote or underserved areas. I'm convinced that future devices must be intuitive and easy to use, even for patients with limited technical expertise.

Furthermore, the investigation of novel applications for acoustic therapy devices warrants attention. While current applications are primarily focused on cardiovascular and respiratory disease, acoustic therapy has the potential to be used in a much wider range of medical conditions. For instance, acoustic stimulation has been shown to have analgesic effects and could potentially be used to manage chronic pain. It may be used in the treatment of neurological disorders, such as Parkinson's disease or Alzheimer's disease. Exploring these novel applications requires conducting preclinical studies to investigate the mechanisms of action of acoustic therapy and conducting clinical trials to evaluate its efficacy and safety in different patient populations.

Finally, ethical considerations surrounding the use of acoustic therapy devices must be addressed. As these devices become more sophisticated and widely used, it is important to ensure that they are used responsibly and ethically. This includes addressing issues such as data privacy, security, and potential biases in algorithms. Researchers must work closely with ethicists, policymakers, and patient advocacy groups to develop guidelines and regulations that promote the responsible use of acoustic therapy devices. This will help to ensure that these devices are used to benefit patients and improve healthcare outcomes, while minimizing the risks of harm or misuse. Think about the security of patient data, and whether or not a cyberattack on a smart device could result in harm.

www.ingramcontent.com/pod-product-compliance
Lightning Source LLC
LaVergne TN
LVHW022230200926
848481LV00006B/2205
9798349520983